RAPID BIOASSESSMENT OF STREAM HEALTH

RAPID BIOASSESSMENT OF STREAM HEALTH

Edited by
Duncan L. Hughes
Michele P. Brossett
James A. Gore
John R. Olson

CRC Press
Taylor & Francis Group
Boca Raton   London   New York

CRC Press is an imprint of the
Taylor & Francis Group, an **informa** business

CRC Press
Taylor & Francis Group
6000 Broken Sound Parkway NW, Suite 300
Boca Raton, FL 33487-2742

© 2010 by Taylor and Francis Group, LLC
CRC Press is an imprint of Taylor & Francis Group, an Informa business

Printed in the United States of America on acid-free paper
10 9 8 7 6 5 4 3 2 1

International Standard Book Number: 978-1-4200-9091-8 (Hardback)

Contents

Preface

Biological assessment is a process whereby knowledge of the condition of an ecological system is obtained by field sampling of targeted biological components of an ecosystem (for freshwater systems, typically benthic macroinvertebrates, fish, or periphyton), data analysis, and comparison to baseline or background conditions. That knowledge is then used to make management decisions regarding ecological restoration and protection, whether in the form of promulgated, legal criteria and standards, or as a means of enhancing public awareness and fostering stewardship. In any case, key to the effectiveness of biological assessments in environmental protection is defensibility, that is, the ability to make statements of condition with known confidence. The more that is known about the quality of the data supporting biological assessments, the greater is their defensibility. This book, *Rapid Bioassessment of Stream Health,* authored by James Gore, Duncan Hughes, Michele Brossett, John Olson, and Gore's students, is a step in that direction for the state of Georgia.

The book results from a Clean Water Act Section 319 grant to Columbus State University, Georgia, which focused on characterizing reference conditions for the state's ecoregions (Levels III and IV) and developing a framework applicable to the fresh, nontidal, lotic surface waters of Georgia. The authors have applied concepts and principles derived from the U.S. Environmental Protection Agency's Rapid Bioassessment Protocols and the multimetric Index of Biological Integrity, customized them to the ecoregions of Georgia, discussed comparability of that with predictive modeling approaches (specifically, the River Invertebrate Prediction and Classification System [RIVPACS]), evaluated various aspects of data quality (including reference site selection, sample size, taxonomic resolution, and selected quality control practices), and presented a case study using the appropriate index to evaluate the effectiveness of an urban best management practice.

As the statewide database for biological assessments grows, there will be an increased understanding of the variability of background (or reference) conditions, and specific numbers defining criteria for evaluations of ecological degradation (that is, biological response) may evolve. Further, it is also widely recognized that results of biological assessments are intended to inform resource managers of relative conditions, that they do not always tell exactly what is causing degradation, and that additional, more detailed analyses or data collection may be required for identifying specific stressors. That this is an iterative process affected by expanding data sets and improved data management systems is an accepted feature of current biological assessment programs in the United States. Consistent application of biological indicators developed using approaches outlined in this book should lead to the advancement of water resources protection in Georgia, and the authors and students are to be commended for that.

James B. Stribling, Ph.D.
Tetra Tech, Inc.
Center for Ecological Sciences
Owings Mills, Maryland

Acknowledgments

Funding for the Georgia Ecoregions Project was provided through a grant from the Georgia Department of Natural Resources, Environmental Protection Division (GAEPD), under the auspices of the U.S. Environmental Protection Agency (specifically, U.S. Environmental Protection Agency Clean Water Act, Section 319(h) FY 98–Element 1) funding. We thank Bill Kennedy, Shannon Winsness, Tim Pugh, and many others in the Watershed Planning and Monitoring Program, GAEPD, for their guidance and assistance during the course of the field and laboratory work and data analysis.

We are most appreciative of the help provided by Dr. Michael Barbour and Dr. James L. Stribling, Tetra Tech, Inc., who trained us in the use of EDAS (Ecological Data Application System) and in multimetric analysis, as well as the general philosophies and protocols for application of the Rapid Bioassessment Protocols.

A large number of faculty, staff, and students from Columbus State University, participated in this program. Dr. Art Cleveland and Dr. Glenn Stokes, Department of Biology, Columbus State University, provided the necessary quality assurance/quality control (QA/QC) oversight for both field and laboratory analysis as well as ensuring that software installation was correct and that extensive data entry into EDAS was accurate and as close to error-free as possible. Rita Snell provided valuable office management and logistical help. Finally, we thank the graduate students Tracy Ferring, Jof Mehaffey, Amanda Middleton, Marcie Parrish, Salini Pillai, Uttam Rai, Ashley Scott, Jodi Williams, and George Williams for their dedication to both field and laboratory work. Gu Cheng, Todd Smith, and Julio Gutierrez spent many long hours completing our in-lab chemical analyses. Staff technicians Susan Nichols and Brigitte Toon worked both in the field and laboratory while undergraduate field assistants, Josh Goodwin, Rawl Hercules, Tom Willits, and many one- and two-trip volunteers made the sampling easier and more comfortable.

Finally, thanks to all of our families and friends for their support, encouragement, and patience during this process.

Editors

Duncan L. Hughes is the Watershed Coordinator for the Soque River Watershed Partnership in Clarkesville, Georgia. The partnership identifies and addresses sediment and bacteria issues related to nonpoint source pollution. He received a B.S. in environmental health from the University of Georgia and an M.S. in environmental science from Columbus State University. Hughes's research interests include the effects of changing land use and land cover on stream biotic communities and water quality.

Address correspondence to: 301 Wildwood Circle, Clarkesville, GA 30523; phone: (706) 754-7872; e-mail: dhughes@northgatech.edu.

James A. Gore, Ph.D., is Professor of Biology and Dean of the College of Natural and Health Sciences at the University of Tampa, Florida. Dr. Gore received his B.A. degree from the University of Colorado and M.A. and Ph.D. degrees (zoology) from the University of Montana. He has held professorships at the University of Tulsa, Oklahoma; Eminent Scholar Chair in Environmental Science in the Troy State University (Alabama) system; Professor and Chair of the Department of Environmental and Health Sciences at Columbus State University, Georgia; and Professor and Chair of the Department of Environmental Science, Policy, and Geography at the University of South Florida, St. Petersburg. He is a Fulbright scholar having held senior research fellowships in Israel and southern Africa. Dr. Gore has more than 135 publication credits including two books, *The Restoration of Rivers and Streams* (Ann Arbor Science, 1985) and *Alternatives in Regulated River Management* (CRC Press, 1989) and more than 75 papers, book chapters, and technical reports in aquatic biology and hydrology. His primary research interests include the influence of channel hydraulics on the distribution of riverine biota, establishing conservation flows for river ecosystems, and the potential impacts of climate change on the success of invasive species.

Address correspondence to: College of National and Health Sciences, University of Tampa, 401 W. Kennedy Blvd., Box V, Tampa FL 33606-1490; phone: (813) 257-3095; fax: (813) 258-7361; e-mail: jgore@ut.edu.

Michele P. Brossett is an environmental specialist with the Department of Natural Resources in the Watershed Protection Branch of the Environmental Protection Division. Brossett received a B.A. degree in environmental soil science and a B.S. in environmental health from the University of Georgia. She received an M.S. in environmental science from Columbus State University. Brossett is involved with biocriteria for the state of Georgia, nutrient criteria, taxonomic identification, and stream assessments. She, along with two coworkers, conducts a macroinvertebrate stream bioassessment workshop. She has been an author on three reports and five presentations.

Address correspondence to: Department of Natural Resources, Environmental Protection Division, Intensive Survey Unit, 4220 International Parkway, Suite

101, Atlanta, GA 30354; phone: (404) 675-1683; fax: (404) 675-6244; e-mail: michele.brossett@dnr.state.ga.us.

John R. Olson is a doctoral candidate in the Watershed Sciences Program of the College of Natural Resources at Utah State University. He received a B.A. degree in geography from the University of California, Santa Barbara, and an M.S. in environmental science from Columbus State University. He has been involved with several state-level bioassessment programs, including ones in Georgia, Arizona, Wyoming, and Utah. Olson has been an author on four journal articles plus more than thirteen presentations, posters, and technical reports on bioassessment and uses of information systems (GIS). He was awarded the Best Student Methods Presentation Award at the 2006 Annual Meeting of the North American Benthological Society, in Anchorage, Alaska. Olson's primary research interests include aquatic ecosystem ecology; bioassessment, spatial and macroecology; and the effects of scale on environmental research.

Address correspondence to: Utah State University, Department of Aquatic, Watershed, and Earth Resources, 5210 Old Main Hill, Logan, UT 84322; phone: (435) 770-4533; fax: (435) 797-1871; e-mail: john.r.olson@aggiemail.usu.edu.

Contributors

James Banning
University of South Florida,
   St. Petersburg
St. Petersburg, Florida

Tracy J. Ferring
Fort Benning Environmental
   Management
Fort Benning, Georgia

Amanda M. Herrit
Aquaterra Engineering, LLC
Chattanooga, Tennessee

Michele de la Rosa
University of South Florida,
   St. Petersburg
St. Petersburg, Florida

Erik Oij
University of South Florida,
   St. Petersburg
St. Petersburg, Florida

Uttam K. Rai
Rhithron Associates, Inc.
Missoula, Montana

Jodi A. Williams
Fort Benning Environmental
   Management
Fort Benning, Georgia

1 Introduction

James A. Gore, Duncan L. Hughes, Michele P. Brossett, and Amanda M. Herrit

CONTENTS

As part of the Clean Water Act (CWA) Section 101(a), it is the obligation of each state to monitor and assess the chemical, physical, and biological conditions of streams within its boundaries. Therefore, states are required to consider the "biological integrity" of their waters when developing stream monitoring procedures (Berry and Dennison 2000). Biotic indices are accepted by the Environmental Protection Agency (EPA) as a method for assessing the biological health or condition of wadeable streams (Barbour et al. 1999).

States must also determine water quality standards for all water bodies as required by CWA Section 305(b). Water quality standards establish designated use and criteria for each water body, which must be maintained for all waters within each state (Berry and Dennison 2000). State agencies must first define water quality standards and then determine a method of monitoring these standards. In some states, biological indices have been used to assess and monitor streams in order to maintain water quality standards set by states (Barbour 1997).

Beginning in the 1970s with the CWA, biological monitoring has developed into a widely used tool for tracking the condition of water resources. In the United States, the chemical condition of water resources was the only consideration in monitoring and remediating processes before the 1970s (Berry and Dennison 2000). During the last 20 years, the United States has made great improvements eliminating point-source pollution and, as a result, chemical contamination has been greatly reduced. Currently, the major impairment concern for surface waters is nonpoint source pollution (Barbour 1997). Biological assessment has been found to be an equally effective tool for assessing both point and nonpoint source pollution (Karr 1991).

For water bodies that have been shown to be impaired, states must develop a plan for returning that water body to an unimpaired status. Important regulatory controls, intended to accomplish this task, are total maximum daily loads (TMDLs) of target nonpoint source contaminants, ranging from metals and nutrients to suspended sediment. According to CWA Section 303(d), state regulatory agencies must establish TMDLs for each water body that has not attained water quality standards after imposing technology-based controls (Barbour et al. 1999). Biological assessments of the structure and function of lotic communities can determine whether or not water

1

quality standards have been achieved and if TMDLs are required for a specific water body.

A broadly applicable indicator for use in biological assessment is the Index of Biotic Integrity (IBI) (Karr 1981). The IBI approach was developed to identify levels of stream impairment using the fish assemblage as a biological indicator. Using the IBI as a model, many biomonitoring programs have expanded to incorporate several types of multimetric indices using fish, macroinvertebrates, and periphyton assemblage-level data. It has been shown that using multiple assemblages from various trophic levels can provide assessments of a broader array of stressors causing stream impairment (Karr 1991).

By monitoring biological indicators, such as the benthic macroinvertebrate community, researchers can describe a given stream or river condition. Macroinvertebrates are considered excellent indicators because they are relatively sedentary and thus can be used to assess long-term change and cumulative effects in a specific location. And, depending upon the number of sampling locations and monitoring network design, biological indicators can be used in broader-scale assessments, such as catchments.

According to Murtaugh (1996), an indicator is considered effective if it is sensitive to stressors or other specific factors being considered. When investigating stream conditions, macroinvertebrate community assemblages can provide researchers with a description of the stream's condition (Resh 1995). Stream community structures are altered by human disturbance and can be used to identify the type and level of disturbance encountered. Using ecological descriptors such as tolerance/intolerance values, macroinvertebrate assemblages can describe the impairment level of a stream relative to the chosen reference condition (Barbour et al. 1999).

The Rapid Bioassessment Protocol (RBP) was developed as a cost-effective and time-efficient procedure for assessing wadeable streams (Barbour et al. 1999). In its most complete, but rarely applied form, the RBP uses fish, benthic macroinvertebrate, and periphyton assemblage-level data to develop a multimetric index, which is used as the indicator of stream impairment. Metrics are used to quantify different attributes of the stream biota (Jessup and Gerritsen 2000). The choice of final metrics ultimately used in an index is based, in part, on their relationship to ecoregional characteristics and response to stressors (Barbour et al. 1999).

Several methods have been used to classify streams according to their abiotic characteristics. Ricker (1934) developed a stream classification for streams in Ontario, Canada, based on the size of the stream, substrate material, the diversity and abundance of the biota, and the physical and chemical characteristics of the water body. Ricker developed this system to group streams according to their similar abiotic characteristics. This technique was used to investigate streams with similar physical and chemical properties, therefore defining each stream's biological characteristics without biased abiotic information.

More recently, Omernik (1987) developed a map of ecoregions for the United States as a framework for grouping ecosystems, both aquatic and terrestrial. Ecoregional groups are based upon patterns of topography, geology and soil, and land use that are intended to minimize within-group variability and maximize among-group variability. For an index to be effective, abiotic differences such as variation in catchment geology must first be eliminated. Using the ecoregion approach, multimetric indices can be more easily calibrated to detect impairment (Jessup and Gerritsen 2000). Therefore, biotic indices are developed specifically for each ecoregion and the streams within are compared to assess each stream's condition (Paul and Gerritsen 2002).

Multimetric indices are used to describe the ecological characteristics and to detect threats to biological integrity of a stream (Rankin 1995). Metrics from richness, composition, tolerance/intolerance, and habit/trophic biological categories are evaluated to determine their ability to detect differences in reference and impaired conditions. Streams are grouped according to their physical and chemical characteristics and are compared within groups. Usually between six to eight metrics are chosen for an index and assigned a quantitative index score for each stream. Based upon variation from the least impaired sites, the index score describes each stream relative to its level of impairment. Once a quantitative rating is assigned, the index score can be described by a qualitative rating.

Using quantitative index scores to describe streams within groups, narrative ratings describe stream characteristics qualitatively. Narrative ratings typically group streams into *good*, *fair*, and *poor* qualitative categories. Each stream is evaluated based on its potential to achieve the least impaired condition within each group. Qualitative measures of stream condition can be used to determine regulatory and monitoring needs of each stream. Using narrative biological criteria, monitoring agencies can determine action plans for stream conservation and restoration (Karr and Chu 1999).

In 1977, Hilsenhoff introduced his biotic index based on organic and nutrient tolerance/intolerance levels of arthropods. Using one order, Hilsenhoff was able to simplify the bioassessment process. Hilsenhoff's biotic index was based upon a 100-individual sample in which each species or genus of arthropod was assigned a tolerance/intolerance level. Once all individuals from each sample were identified, the tolerance/intolerance values were averaged together giving each stream a biotic index score (Hilsenhoff 1987).

Originally, the CWA standard for adequate biological support was termed "fishable-swimmable" but this standard has evolved into a more functional "aquatic life use" designation (Berry and Dennison 2000). Multimetric indices can also be used to determine aquatic life use designations, an EPA requirement for nonpoint source management. The Vermont Department of Environmental Conservation has employed benthic and fish data, by means of multimetric indices, to determine numeric biological criteria. Numeric criteria are applied and are used to evaluate each water body according to aquatic life use designations. Being the quantitative equivalent of narrative biological criteria, numeric biological criteria can also be used to assess water quality standards (Vermont Department of Environmental Conservation 2004).

Several states have developed narrative rating systems to describe numeric biological criteria. In Ohio, the Qualitative Habitat Evaluation Index (QHEI) was developed to determine the aquatic life potential of each water body. Each water body is assigned an aquatic use level, which could be applied to aquatic life use designations. The purpose of this system was to describe the physical, chemical, and biological properties of a water system, and therefore protect all facets of the system (Rankin

1989). Narrative criteria are used to describe the water body's condition or current state, which is based on quantitative data. The QHEI has two main categories of aquatic life uses: warm water habitat (WWH) and exceptional warm water habitat (EWH). The WWH is described as the typical habitat condition of rivers and streams in Ohio. The EWH is an aquatic habitat that is exceptional for its fauna and quality of habitat. Narrative criteria of exceptional (EWH), good (WWH), fair, poor, or very poor are assigned to each stream or river (Rankin 1989).

The Benthic Index of Stream Integrity (BISI), developed in Rockdale County, Georgia, assigns each stream with a quantitative rating. Using a percentile method, the index score is described by a qualitative rating. Streams with an index score above the 25th percentile are equally divided into *good* and *very good* narrative ratings. Streams rated below the 25th percentile are divided into three groups: *fair*, *poor*, and *very poor*. Narrative ratings are used to describe biological characteristics that are found in each stream category (Tetra Tech 2001).

With the use of multimetric indices, chemical analysis, and physical habitat assessment, stream assessment methods have been developed to identify the level of stream impairment. Once stream assessment is completed, this information can be used to determine regulatory and monitoring procedures for the study area. The evaluation of stream conditions is an important method for managing water resources (Barbour et al. 1999).

In this book, we demonstrate the application of the EPA's RBP to fit the needs of the regulatory agency governing the health of stream and river ecosystems. Although our experiences were in the development and application of this index for use in the state of Georgia and our examples, necessarily, will come from that project, we feel that the system we employed is appropriate for any region in the world.

In 1996, the Georgia Environmental Protection Division (GAEPD) began a multiphase project to develop biological criteria for wadeable streams. Phase I, the Georgia Ecoregion/Subecoregion Delineation and Reference Site Selection, was the initial step for biological criteria development. The primary objective of Phase I was to develop a useful, general purpose, geographical framework that categorizes large sections of Georgia into logical units of similar geology, physiography, soils, vegetation, land use/land cover, and water quality. The key output of this project was to refine the Level III ecoregions (delineated by Omernik in 1987) and then to subdivide these refined Level III ecoregions into Level IV ecoregions. In June 2000 a draft map was produced and in 2001 a final map was produced that showed Level IV ecoregions for Georgia as well as potential locations of candidate reference sites for the next phase of the multiphase project (Griffith et al. 2001). Phase II focused on (1) developing land use judgment criteria for candidate reference sites in Level IV ecoregions that were delineated in Phase I and (2) conducting an intensive sampling of the reference sites. Characterization of resident biota inhabiting those reference sites established baseline, best attainable reference conditions representative of each Level IV ecoregion. The objective of this study was to collect and analyze chemical and biological water quality samples at reference sites that are representative of Level IV ecoregions across the state. Phase III of this project focused on evaluation of impaired streams in comparison to reference characteristics in each of the major ecoregions and subecoregions in Georgia. Specific activities included the identification of a suite of impaired sites in each of the ecoregions and distinctive subecoregions (identified in Phase II), which were sampled, using RBP techniques, for physical, chemical, and biological condition. A random subset of reference streams (from Phase II) was also resampled as a validation and verification of the reference characterization. As in Phase II, the impaired streams were identified using unbiased geographic information system (GIS) analysis that identified classes of impaired streams according to various land use patterns. Using multivariate techniques, perfected in Phase II, these impacted sites were analyzed to characterize the level of disturbance and the assessment criteria that defined that level. The objective of Phase III was to identify trends and establish a numerical scoring system (i.e., biological criteria for macroinvertebrates) and to validate the results. The objectives of Phase IV were the verification and validation of the numerical scoring system, as well as development of a defensible system for applying the numerical scoring system to evaluate the health of other streams in Georgia. In addition, a framework for the application of bioassessment to various regulatory activities (such as TMDLs or other CWA Section 303(d) requirements) was developed.

Using RBP techniques, we described the reference and impaired condition of streams in Georgia. For quality control, a random selection of reference sites was resampled to verify the findings. Both reference sites and impaired sites were selected via a unique unbiased approach using GIS land use data. Using a multimetric approach to scoring the macroinvertebrates living in the selected reference and impaired streams, we produced a numeric scoring system comparing streams within an ecoregion (or subecoregion) that can be used to assess current ecosystem status and to identify trends after development or other alteration of the catchment. By the use of multimetric indices, the rating system designates a numeric value for each stream. This numeric value can be used to determine regulatory action for each stream within an ecoregion. The stream rating system incorporates benthic macroinvertebrate, chemical and physical habitat data to produce a robust assessment tool. The assessment system will supplement previous assessment methods from other states (Mississippi Department of Environmental Quality [MDEQ] 2003) by providing a method for not only assessing streams but also regulating compliance with water quality regulations for streams in ecoregions of Georgia.

We hope that this book will guide other aquatic scientists and water resource managers in making the critical decisions to implement or modify our techniques to develop a similar assessment system for their states or regions.

REFERENCES

Barbour, M.T. 1997. The re-invention of biological assessment in the United States. *Human and Ecological Risk Assessment* 3: 933–940.

Barbour, M.T., J. Gerritsen, B.D. Snyder, and J.B. Stribling. 1999. *Rapid Bioassessment Protocols for Use in Streams and Wadeable Rivers: Periphyton, Benthic Macroinvertebrates and Fish*, 2nd ed. EPA 841-B-99-002. Washington, DC: U.S. Environmental Protection Agency, Office of Water.

Berry, J.F., and M.S. Dennison. 2000. *The Environmental Law and Compliance Handbook.* New York: McGraw-Hill.

Griffith, G.E., J.M. Omernik, J.A. Cornstock, S. Lawrence, and T. Foster. 2001. *Level III and IV Ecoregions of Georgia* [ecoregional boundary data sets in a polygonal vector format as ArcInfo export coverage on the Internet]. Revision 5. Corvallis, OR: U.S. EPA, National Health and Environmental Effects Research Lab/ORD, Western Ecology Division. (Can be obtained from Georgia Department of Natural Resources, Atlanta.)

Hilsenhoff, W.L. 1987. An improved biotic index of organic stream pollution. *The Great Lakes Entomologist* 20: 31–39.

Jessup, B.K., and J. Gerritsen. 2000. *Development of Multimetric Index for Biological Assessment of Idaho Streams using Benthic Macroinvertebrates*. Prepared for the Idaho Department of Environmental Quality. Owings Mills, MD: Tetra Tech, Inc.

Karr, J.R. 1981. Assessment of biotic integrity using fish communities. *Fisheries* 6: 21–27.

Karr, J.R. 1991. Biological integrity: A long-neglected aspect of water resource management. *Ecological Applications* 1: 66–84.

Karr, J.R., and E.W. Chu. 1999. *Restoring Life in Running Waters: Better Biological Monitoring*. Washington DC: Island Press.

Mississippi Department of Environmental Quality (MDEQ). 2003. *Development and Application of the Mississippi–Benthic Index of Stream Quality (M-BISQ).* Jackson: Mississippi Department of Environmental Quality, Office of Pollution Control.

Murtaugh, P.A. 1996. The statistical evaluation of ecological indicators. *Ecological Applications* 6: 132–139.

Omernik, J.M. 1987. Map supplement: Ecoregions of the conterminous United States. *Annals of the Association of American Geographers* 77: 118–125.

Paul, M.J., and J. Gerritsen. 2002. *Draft Statistical Guidance for Developing Indicators for Rivers and Streams: A Guide for Constructing Multimetric and Multivariate Predictive Bioassessment Models.* Cincinnatti, OH: EPA Office of Research and Development.

Rankin, E.T. 1989. *The Qualitative Habitat Evaluation Index (QHEI): Rational, Methods, and Application.* Columbus: State of Ohio Environmental Protection Agency, Ecological Assessment Section, Division of Water Quality, Planning, and Assessment.

Rankin, E.T. 1995. Habitat indices in water resource quality assessment. In *Biological Assessment and Criteria: Tools for Water Resource Planning and Decision Making*, eds. W.S. Davis and T.P. Simon, 181–208. New York: Lewis Publishers.

Resh, V.H. 1995. Freshwater benthic macroinvertebrates and rapid assessment procedures for water quality monitoring in developing and newly industrialized countries. In *Biological Assessment and Criteria: Tools for Water Resource Planning and Decision Making*, eds. W.S. Davis and T.P. Simon, 167–177. New York: Lewis Publishers.

Ricker, W.E. 1934. *An Ecological Classification of Certain Ontario Streams* (University of Toronto Press, Publications of the Ontario Fisheries Research Laboratory, No. 49, University of Toronto Studies, Biological Series No. 37). Toronto: University of Toronto Press.

Tetra Tech, Inc. 2001. *Watershed Characterization Report*. Rockdale County Report. Rockdale, GA.

Vermont Department of Environmental Conservation (VDEC). 2004. *Biocriteria for Fish and Macroinvertebrate Assemblages in Vermont Wadeable Streams and Rivers: Development Phase.* Burlington, VT: VDEC, Water Quality Division, Biomonitoring and Aquatic Studies Session.

2 Comparison of Bioassessment Methods

John R. Olson, Duncan L. Hughes, and Michele P. Brossett

CONTENTS

INTRODUCTION

Various bioassessment methods are available to resource managers and investigators to gauge biological integrity of lotic systems, in addition to the multimetric approach used in the Rapid Bioassessment Protocol (RBP). Although the RBP and its associated multimetric approach are widely used in the United States, other protocols to evaluate biotic heath are preferred elsewhere (Diamond et al. 1996). Predictive models to assess biological health and ecological integrity (also know as multivariate or River Invertebrate Prediction and Classification System [RIVPACS] type models) are commonly used in Europe, Australia, parts of the United States, and Canada. In addition to these two approaches, many other methods of assessing water quality using benthic macroinvertebrates exist (see Merritt et al. 2008, for a listing). However, none of these methods are as widespread in their use as the predictive model and multimetric approaches, and we therefore will limit our discussion to only these two approaches.

This text is not intended to provide an exhaustive discussion of the methodology of these two approaches or their relative strengths and weaknesses. Rather, we will briefly introduce the predictive modeling approach to bioassessment using the RIVPACS framework, and describe some of the similarities and differences between the multimetric and predictive modeling methods. A comparison of selected bioassessment approaches may be found in Table 2.1.

TABLE 2.1

Comparison of Selected Bioassessment Approaches

	Multimetric Approach	Predictive Model Approach	
Item	RBP	RIVPACS	AusRivAS
Reference Site Selection	Ecoregional	Combined local and catchment attributes	Combined local and catchment attributes
Reference Site Criteria	Minimally impaired; BPJ or anthropogenic variables	Minimally impaired; BPJ or anthropogenic variables	Minimally impaired; BPJ or anthropogenic variables
Habitat Sampled	Single or multiple	Single or multiple	Single or multiple
Sample Protocol	Composite sample, equal effort	Composite sample, equal area or effort	Composite sample, equal area or effort
Taxonomic Resolution	Lowest practicable, all taxa used	Lowest practicable, all taxa used	Family, only common taxa used
Date Type	Abundance	Presence/absence	Presence/absence
Control of Natural Gradients	*A priori* partitioned into ecoregions	Using continuous variables in discriminant model	Using continuous variables in discriminant model
Analysis Requirements	Spreadsheet	Cluster analysis, discriminant model, and spreadsheet	Cluster analysis, discriminant model, and spreadsheet
Output	Combined multimetric score for each ecoregion	Ratio of observed number of taxa to expected	Ratio of observed number of taxa to expected

Note: BPJ = Best professional judgment.

THE PREDICTIVE MODELING APPROACH TO BIOASSESSMENT

All predictive modeling approaches to bioassessment are based upon or share key principles with the RIVPACS predictive model developed at the Institute of Freshwater Ecology in the United Kingdom (Wright 1995). These include primarily AusRivAS (Australian River Assessment System) and BEAST (Benthic Assessment of Sediment), initially used in Canada exclusively in the Great Lakes but later adapted for streams and rivers (Reynoldson et al. 1995, 2001).

The predictive modeling approach makes no *a priori* assumptions about macroinvertebrate communities based on physiochemical or habitat data. Rather, reference sites are classified into biologic groups with similar community composition. A statistical model based upon the physiochemical and habitat data is then used to predict the probability of a new site (one being tested for impairment) belonging to each one of these groups. The known frequency of occurrence of each taxon within each biologic group is then combined with the probability of a new site belonging to each group to produce an expected number of taxa at the new site. The benthic taxa observed (O) at each test site is then compared to expected taxa (E), in the form of a ratio (O/E). An O/E ratio within the model's error of 1 indicates the system is unimpaired. The further the O/E ratio deviates from 1, the greater the level of impairment (Australian Government Department of Environment and Heritage [AGDEH] 2004). The RIVPACS predictive modeling approach, upon which both AusRivAS and BEAST are based, employs a multistep process. The general framework involves the following steps (Hawkins and Carlisle 2001; Herbst and Silldorff 2006):

1. Classify all reference sites into biologically similar groups (most frequently done using cluster analysis and excluding the rarest taxa).
2. Develop decision rules for classifying new test sites into the groups identified in Step 1 based on measured characteristics of the stream and its watershed that are not affected by anthropogenic stress (typically accomplished with a discriminant model).
3. Use the decision rules established in Step 2 to assign the probability of sites belonging to each of the groups identified in Step 1 (typically obtained through the discriminant analysis routine or software).
4. Calculate the probability that each taxon will be captured within each of the groups identified in Step 1 (calculated by dividing the number of sites in each group that had a particular taxon by the total number of sites within the group).
5. Calculate the probabilities that each taxon will occur at a new site by weighting the probability of the taxon being captured within a group by the probability of the new site belonging to the group (by multiplying the results of Step 3 and Step 4, noting that a new site often has a probability of occurring in more than one group).
6. Calculate the expected taxa richness (E) as the sum of the probabilities from Step 5 (usually only including taxa with an E > 0.5) and the observed taxa richness (O) of a sample from the new site, and then calculate the ratio of these values (O/E) to create an index.

The O/E ratio represents the number of taxa still found at the site, compared with what that site is expected to have in its natural condition. New sites with O/E ratios that score lower than the model error around 1.0 are considered to be affected by anthropogenic stressors (Herbst and Silldorff 2006).

While both RIVPACS and AusRivAS attempt to predict benthic communities (in the absence of disturbance) at new sites using taxa expected from similar reference streams, the AusRivAS method and U.S. applications of the RIVPACS approach differ from RIVPACS in that they calculate an O/E ratio only for common taxa having probabilities of capture of >0.5. AusRivAS also only identifies macroinvertebrates to family level, eliminating some laboratory processing time, and uses invertebrates sampled and processed from each major habitat separately (AGDEH 2004).

SIMILARITIES BETWEEN MULTIMETRIC
AND PREDICTIVE MODELING APPROACHES

Although the multimetric and predictive model approaches are used in similar ways for bioassessment and are based upon the idea that biota are indicators of stream health, they were originally developed from different perspectives. The multimetric approach was originally designed as a way of quantifying multiple characteristics of fish assemblages in the Midwest (United States). The characteristics chosen were those known to be sensitive to human-caused disturbances and these were then combined to produce an overall measure of biotic integrity as put forth by Frey (1977), Karr and Chu (2000), and Whittier et al. (2007). The predictive modeling approach was originally developed to produce biological classifications of unimpaired streams in Great Britain, based upon benthic macroinvertebrates fauna and to predict biological class membership from physical and chemical features (Wright et al. 1998). However, as these two approaches have evolved over the years, they now have become similar in many respects.

Both approaches have now been applied to a wide range of biotic assemblages, including periphyton, macroinvertebrates, and fish in lotic, lentic, and wetland systems. There are not any appreciable differences in how biota are collected in the field and both approaches can be used to assess samples taken from either single or multiple habitats within a stream. Sample processing methods and the levels of taxonomic resolution used to identify the biota are also now similar in both approaches. Generally, as taxonomic resolution, sample size, and repeatability of biotic data collected improve, the precision and ability of each of these approaches to detect impairment improves. Since both approaches use similar input data, single data sets are now analyzed with both approaches on occasion (e.g., Herbst and Silldorff 2006; Stribling et al. 2008).

Each approach also relies upon the selection of representative, minimally impaired reference streams, with which test sites are compared. The need for a minimally impaired biological baseline is vital in determining the degree and nature of stream impairment across a disturbance gradient (Reynoldson et al. 1997). Also, biota found in reference streams are used in all programs to set realistically attainable biological goals for impaired streams.

DIFFERENCES BETWEEN MULTIMETRIC
AND PREDICTIVE MODELING APPROACHES

The two major differences between multimetric and predictive modeling approaches is in how they account for naturally occurring differences among streams and the number of biological signals they use in assessing overall stream biotic condition (Karr and Chu 2000; Norris and Hawkins 2000). Predictive models summarize changes in stream biotic condition only in terms of taxonomic completeness, the O/E ratio. Although multimetric indices do not include a measure of completeness, they do incorporate measures of composition by different taxa. In addition, multimetric indices use measures of habitat, tolerance, trophic levels, life history, abundance, presence of invasive taxa, and presence of disease or deformities among assemblages. Where multimetric approaches primarily rely on *a priori* classifications of the landscape into regions to partition natural variation, predictive models treat natural variation as continuous gradients without necessarily referencing spatial patterns. The multimetric approach also minimizes the effect of environmental gradients by selecting metrics that are minimally affected by natural gradients but respond maximally to human disturbance.

Because streams can vary quite dramatically across the landscape and the biota at any site will reflect that stream's characteristics, it is important that bioassessment methods have a way of accounting for naturally occurring variation. The multimetric approach's reliance on spatial classifications to partition natural variance assumes that the natural variation among streams shows a somewhat spatially homogeneous pattern allowing variation to be adequately characterized by spatial classes. Ecoregions (Omernik 1987) are the primary approach to controlling for the effects of natural variation in multimetric indices (although see Stoddard et al. 2008 or Whittier et al. 2007 for examples of calibrating metrics to environmental gradients). Although these ecoregions can be modified to better match patterns of macroinvertebrate distributions (Barbour et al. 1996; Gerritsen et al. 2000), this still results in partitioning environmental variation into spatial classes. In areas where the major factors affecting macroinvertebrate assemblages, such as variations in climate, geology, and hydrology, are relatively homogeneous within an area and these areas have sharply defined boundaries, ecoregional classifications may be expected to perform reasonably well. Our work in Georgia provides an example of ecoregions effectively partitioning natural variance, as has been shown for ecoregions across the southeast United States (Feminella 2000). Both Harding et al. (1997) and Robinson and Minshall (1998) showed examples of ecoregions effectively partitioning natural variation where the major environmental gradients are coincident with one another. Heino and colleagues (2002) showed a similar partitioning of variation across Scandinavia, but noted that this was weaker than expected possibly due to the continuous nature of the variation in benthic invertebrate assemblages. As natural variation becomes more continuous, as opposed to grouped in distinct regions associated with diverse topographies, spatial classes will be less able to capture the variation. However, if the important environmental factors are spatially heterogeneous, vary in continuous spatial gradients instead of discrete areas, or if the major gradients run orthogonal to one another, then the predictive model approach should be expected to perform better because of its non-geographic approach. This is the situation in the western United States, where the geologic and physiographic complexities are difficult to capture in an ecoregional framework and most major physiographic, climatic, and geologic gradients associated with the major mountain ranges run east–west, whereas the major latitudinal temperature gradient runs north–south. Van Sickle and Hughes (2000) showed that ecoregions in western Oregon only weakly partitioned natural variation. Hawkins and Vinson (2000) showed similar results when applying ecoregions across the western United States, even when the regions were created post hoc. Waite et al. (2000) showed a similar example from the Mid-Atlantic Highland region of the Appalachians, where physiographic patterns trend east–west and temperatures trend north–south. In Europe, predictive models also were shown to better partition variation for sites across Britain, Sweden, and the Czech Republic (Davy-Bowker et al. 2006). For a series of papers evaluating the effectiveness of using spatial classifications to account for environmental variation, we direct readers to the *Journal of the North American Benthological Society* (Hawkins et al. 2000). Several researchers have applied models of the natural environment to multimetric indices to better control for natural variability, resulting in improved performance of these indices (Hawkins 2006; Cao et al. 2007; and Pont et al. 2009).

Whereas predictive models focus on community composition, multimetric indices combine information from both community composition and other measurements of biologic integrity. This difference becomes important when conducting bioassessments of streams with naturally low taxa richness, as found in macroinvertebrate assemblages in blackwater streams of the southeastern U.S. coastal plain or in fish assemblages in the western United States. In both of these situations, natural taxa richness is often below 12 species at a site. Because predictive models assess biotic integrity only in terms of the number of naturally occurring taxa at a site, a single chance absence of a taxon results in a larger change in the O/E ratio, as E gets smaller. It has been suggested to minimize the problem that predictive models should only be used where at least 15 taxa are expected (Marchant et al. 2006). However, both Kennard et al. (2006) and Joy and Death (2002) managed to produce predictive models that successfully detected impairment from assemblages with only an expected average of 6 and 5 taxa, respectively. Multimetric assessments, because of their ability to use multiple types of biologic information, do have a distinct advantage in areas of low taxa richness. This is especially true when metrics of individual health or invasive species are used to supplement composition metrics. However, most of the metrics commonly assessed when creating multimetric indices are variations of composition, and therefore should suffer the same weaknesses as predictive models in low richness situations.

Although less well evaluated, the two approaches also differ greatly in how they determine if a site is impaired. The predictive model approach, assessing any site with less than the expected number of taxa expected as impaired, determines impairment independent of any actual measured impairment. In the multimetric approach, both the choice of metrics and the determination of what metric level indicates impairment are done by comparing the response of known impaired sites to unimpaired sites. This approach allows individual metrics to be used to determine the causes of impairment and ensures that metrics that are the most sensitive to impairment are used. However, because these indices are developed to measure specific forms of impairment, they may be less sensitive to forms of impairment not used during metric development. This is especially problematic when new forms of impairment occur, such as industrial accidents, changes in temperature associated with global warming, or invasions by new exotic taxa.

Another difference is the biological and statistical complexity of the analysis used in each approach. Predictive models are much more statistically demanding than the multimetric approach, because of their reliance on cluster analysis to produce biotic classes and discriminant analysis to produce a model for predicting new group membership based on environment variables. Because the performance of predictive models is based upon their precision, and their precision is at least partly a function of the sample size of data used to create the models, the predictive modeling approach is perceived as requiring more data than the multimetric approach. However, a successful predictive model has been created from only 16 reference sites (Hawkins and Carlisle 2001). The multimetric approach, because of its reliance on multiple ecological aspects of the assemblage, requires a higher level of ecological understanding and sophistication than predictive modeling. Both of these forms of complexity impact how the results of the bioassessment are communicated to managers and the public. Although both approaches produce a single number, each number represents something different. The predictive model approach's number is the proportion of naturally occurring taxa that are still present (Hawkins 2006), whereas the multimetric index approach's number is a combination of multiple indicators of biological condition (Karr and Chu 2000). The results from predictive models are simpler to understand, although the process that produces these results is not. Conversely, the process of the multimetric approach is easier to comprehend, but realizing the importance of the results requires a deeper ecological understanding.

CONCLUSION

In spite of these differences, side-by-side comparisons of multimetric and predictive model approaches have shown little difference in their ability to detect impairment. Stribling et al. (2008) evaluated the precision of both approaches using repeated samples from the same location. They found that the relative percent difference between repeated samples varied by about 1% between the methods. A comparison study using both multimetric and predictive modeling approaches at 40 sites in California concluded that similar management decisions can be made using either approach, even with varied field collection or taxonomic protocols (Herbst and Silldorff 2006). The researchers concluded that these results indicate the possibility of comparing outputs among the varied approaches and even integrating results among them to "increase assessment certainty" (Herbst and Silldorff 2006, p. 1277). Lucke and Johnson (2009) reached similar conclusions when they compared multimetric and predictive models in southern Sweden. Hawkins's (2006) comparison of predictive models to multimetric indices also concluded, "O/E assessments resulted in very similar estimates of mean regional conditions compared with most other indicators once these indicators' values were standardized relative to reference-site means." However, he noted that since multimetric indices are built largely on metrics that evaluate change in community composition, the same as O/E, similar results are unsurprising. In the Hawkins study though, ~25% of the sites evaluated resulted in contrary determinations of whether they were impaired. Hawkins (2006) attributed these differences to differences in precision, in sensitivity among indicators to natural environmental variability, and in sensitivity to different stressors. Ultimately, it is difficult to determine which method does a better job at detecting impairment, since the true biotic integrity of a stream can almost never be known. To address this, Cao and Hawkins (2005) used a simulated assemblage to show that the commonly used metric of richness can be biased when used to assess different levels of impairment.

The selection of the multimetric or predictive modeling approach depends largely on the purpose of a biological sampling program, the amount of environmental and biotic variability in the area to be assessed, the technical and financial resources available, and, to some extent, the individual preference of the investigator. Both the multimetric and predictive modeling approaches to biological assessment have been demonstrated to accurately identify test sites as belonging to either the reference or impaired class of streams. However, there are also weaknesses, real or perceived, with both the multimetric (Reynoldson et al. 1997) and predictive modeling methods (Fore et al. 1996).

When choosing between the predictive model and multimetric approaches, the user should consider how the differences outlined earlier might affect the precision and the accuracy of their results. Where biotic richness is naturally low and the environment either changes abruptly or has all of its major gradients coincident, then multimetric indices may have equal or greater accuracy and precision. Where the environment is more heterogeneous, or types of impairment are either unknown or too numerous to easily quantify, then the predictive model approach might work better. Of major concern with the predictive modeling method is the specialization required to develop predictive models and the question of how easily resource managers can use the data to make resource management decisions (Gerritsen 1995). However, developments of user-friendly software, and the lack of a need for resource managers to perform the predictive modeling analysis themselves, allay this fear for some (Reynoldson et al. 1997).

REFERENCES

Australian Government Department of Environment and Heritage (AGDEH). 2004. *National River Health Program AusRivAS Quality Assurance and Quality Control Project. Appendix B: Literature Review–QA/QC Methodology for Rapid Bioassessment Programs.* Canberra, Australia: Department of the Environment, Water Heritage, and Arts.

Barbour, M.T., J. Gerritsen, G.E. Griffith, R. Frydenborg, E. McCarron, and J.W. White. 1996. A framework for biological criteria for Florida streams using benthic macroinvertebates. *Journal of the North American Benthological Society* 15: 185–211.

Cao, Y., and C.P. Hawkins. 2005. Simulating biological impairment to evaluate the accuracy of ecological indicators. *Journal of Applied Ecology* 42: 954–965.

Cao, Y., C.P. Hawkins, J. Olson, and M.A. Kosterman. 2007. Modeling natural environmental gradients improves the accuracy and precision of diatom-based indicators. *Journal of the North American Benthological Society* 26(3): 566–585.

Davy-Bowker, J., R.T. Clarke, R.K. Johnson, J. Kokes, J.F. Murphy, and S. Zahradkova. 2006. A comparison of the European Water Framework Directive physical typology and RIVPACS-type models as alternative methods of establishing reference conditions for benthic macroinvertebrates. *Hydrobiologia* 566: 91–105.

Diamond, J.M., M.T. Barbour, and J.B. Stribling. 1996. Characterizing and comparing bioassessment methods and their results: A perspective. *Journal of the North American Benthological Society* 15(4): 713–727.

Feminella, J.W. 2000. Correspondence between stream macroinvertebrate assemblages and four ecoregions of the southeastern United States. *Journal of the North American Benthological Society* 19(3): 442–461.

Fore, L.S., J.R. Karr, and R.W. Wisseman. 1996. Assessing invertebrate responses to human activities: Evaluating alternative approaches. *Journal of the North American Benthological Society* 15: 212–231.

Frey, D.G. 1977. Biological integrity of waters: An historical approach. In *The Integrity of Water: A Symposium*, eds. R.K. Ballentine and L.J. Guarraia, 127–140. Washington, DC: U.S. Environmental Protection Agency.

Gerrittsen, J. 1995. Additive biological indices for resource management. *Journal of the North American Benthological Society* 14: 451–457.

Gerritsen, J., M.T. Barbour, and K. King. 2000. Apples, oranges, and ecoregions: On determining pattern in aquatic assemblages. *Journal of the North American Benthological Society* 19(3): 487–496.

Harding, J.S., M.J. Winterbourn, and W.F. McDiffett. 1997. Stream faunas and ecoregions in South Island, New Zealand: Do they correspond? *Archiv Fur Hydrobiologie* 140(3): 289–307.

Hawkins, C.P. 2006. Quantifying biological integrity by taxonomic completeness: Its utility in regional and global assessments. *Ecological Applications* 16(4): 1277–1294.

Hawkins, C.P., and D.M. Carlisle. 2001. Use of predictive models for assessing the biological integrity of wetlands and other aquatic habitats. In *Bioassessment and Management of North American Freshwater Wetland*, eds. R. Rader, D. Batzer, and S. Wissinger, 59–83. New York: John Wiley & Sons.

Hawkins, C.P., R.H. Norris, J. Gerritsen, R.M. Hughes, S.K. Jackson, R.K. Johnson, and R.J. Stevenson. 2000. Evaluation of the use of landscape classifications for the prediction of freshwater biota: Synthesis and recommendations. *Journal of the North American Benthological Society* 19(3): 541–556.

Hawkins, C.P., and M.R. Vinson. 2000. Weak correspondence between landscape classifications and stream invertebrate assemblages: Implications for bioassessment. *Journal of the North American Benthological Society* 19(3): 501–517.

Heino, J., T. Muotka, R. Paavola, H. Hamalainen, and E. Koskenniemi. 2002. Correspondence between regional delineations and spatial patterns in macroinvertebrate assemblages of boreal headwater streams. *Journal of the North American Benthological Society* 21(3): 397–413.

Herbst, D.B., and E.L. Silldorff. 2006. Comparison of the performance of different bioassessment methods: Similar evaluations of biotic integrity from separate programs and procedures. *Journal of the North American Benthological Society* 25(2): 513–530.

Joy, M.K., and R.G. Death. 2002. Predictive modelling of freshwater fish as a biomonitoring tool in New Zealand. *Freshwater Biology* 47: 2261–2275.

Karr, J.R., and E.W. Chu. 2000. Sustaining living rivers. *Hydrobiologia* 422/423: 1–14.

Kennard, M.J., B.J. Pusey, A.H. Arthington, B.D. Harch, and S.J. Mackay. 2006. Development and application of a predictive model of freshwater fish assemblage composition to evaluate river health in eastern Australia. *Hydrobiologia* 572: 33–57.

Lucke, J.D., and R.K. Johnson. 2009. Detection of ecological change in stream macroinvertebrate assemblages using single metric, multimetric or multivariate approaches. *Ecological Indicators* 9(4): 659–669.

Marchant, R., R.H. Norris, and A. Milligan. 2006. Evaluation and application of methods for biological assessment of streams: Summary of papers. *Hydrobiologia* 572: 1–7.

Merritt, R.W., K.W. Cummins, and M.B. Berg, eds. 2008. *An Introduction to the Aquatic Insects of North America*, 4th ed. Dubuque, IA: Kendall/Hunt.

Norris, R.H., and C.P. Hawkins. 2000. Monitoring river health. *Hydrobiologia* 435: 5–17.

Omernik, J.M. 1987. Ecoregions of the conterminous United States. *Annals of the Association of American Geographers* 77(1): 118–125.

Pont, D., R.M. Hughes, T.R. Whittier, and S. Schmutz. 2009. A predictive index of biotic integrity model for aquatic-vertebrate assemblages of western U.S. Streams. *Transactions of the American Fisheries Society* 138: 292–305.

Reynoldson, T.B., R.C. Bailey, K. Day, and R.H. Norris. 1995. Biological guidelines for freshwater sediment based on Benthic Assessment of Sediment (the BEAST) using a multivariate approach for predicting biological state. *Australian Journal of Ecology* 20: 198–219.

Reynoldson, T.B., R.H. Norris, V.H. Resh, K.E. Day, and D.M. Rosenberg. 1997. The reference condition: A comparison of multimetric and multivariate approaches to assess water quality impairment using benthic macroinvertebrates. *Journal of the North American Benthological Society* 16(4): 833–852.

Reynoldson, T.B., D.M. Rosenberg, and V.H. Resh. 2001. Comparison of models predicting invertebrate assemblages for biomonitoring in the Fraser River catchment, British Columbia. *Canadian Journal of Fisheries and Aquatic Sciences* 58: 1395–1410.

Robinson, C.T., and G.W. Minshall. 1998. Regional assessment of wadable streams in Idaho, USA. *Great Basin Naturalist* 58(1): 54–65.

Stribling, J.B., B.K. Jessup, and D.L. Feldman. 2008. Precision of benthic macroinvertebrate indicators of stream condition in Montana. *Journal of the North American Benthological Society* 27(1): 58–67.

Stoddard, J.L., A.T. Herlihy, D.V. Peck, R.M. Hughes, T.R. Whittier, and E. Tarquinio. 2008. A process for creating multimetric indices for large-scale aquatic surveys. *Journal of the North American Benthological Society* 27(4): 878–891.

Van Sickle, J., and R.M. Hughes. 2000. Classification strengths of ecoregions, catchments, and geographic clusters for aquatic vertebrates in Oregon. *Journal of the North American Benthological Society* 19(3): 370–384.

Waite, I.R., A.T. Herlihy, D.P. Larsen, and D.J. Klemm. 2000. Comparing strengths of geographic and nongeographic classifications of stream benthic macroinvertebrates in the Mid-Atlantic Highlands, United States. *Journal of the North American Benthological Society* 19(3): 429–441.

Whittier, T.R., R.M. Hughes, J.L. Stoddard, G.A. Lomnicky, D.V. Peck, and A.T. Herlihy. 2007. A structured approach for developing indices of biotic integrity: Three examples from streams and rivers in the western USA. *Transactions of the American Fisheries Society* 136: 718–735.

Wright, J.F. 1995. Development and use of a system for predicting the macroinvertebrate fauna in flowing waters. *Australian Journal of Ecology* 29: 181–197.

Wright, J.F., M.T. Furse, and D. Moss. 1998. River classification using invertebrates: RIVPACS applications. *Aquatic Conservation: Marine and Freshwater Ecosystems* 8: 617–631.

3 Rapid Bioassessment Materials and Methods

Michele P. Brossett, Duncan L. Hughes, John R. Olson, and James A. Gore

CONTENTS

RAPID BIOASSESSMENT PROTOCOL

The Environmental Protection Agency (EPA) has published guidelines on developing bio-assessments, *Rapid Bioassessment Protocols for Use in Streams and Wadeable Rivers: Periphyton, Benthic Macroinvertebrates, and Fish* (Barbour et al. 1999); and biocriteria, *Biological Criteria: Technical Guidance for Streams and Small Rivers* (Gibson et al. 1996). The Rapid Bioassessment Protocol (RBP) is a guide for conducting cost-effective biological assessments of lotic systems. The RBP is an integrated assessment, which compares habitat, water quality, and biological measures to define a reference condition. Bioassessments, and their resulting biocriteria, are effective methods for assessing water quality because they provide a way of integrating the chemical, physical, and biological effects on water quality by directly measuring biologic integrity and indirectly measuring physical and chemical integrity. This indirect measurement of chemical integrity may actually be more representative of the true integrity of the stream than base flow chemical data since chemical composition varies widely over time, both seasonally and during storm flows (Bolstad and Swank 1997; Johnson et al. 1997).

The Assessment and Watershed Protection Division of the EPA, developed the RBP document to provide basic aquatic life data for water quality management purposes such as problem screening, site ranking, and trend monitoring (Barbour et al. 1999). States have found these protocols to be useful as a framework for their biocriteria monitoring programs. The RBP document was meant as a self-corrective process as science advances; the implementation by state water resource agencies has contributed to refinement of the original procedure for regional specificity. The RBP document, revisited in 1999, reflects the advancement in bioassessment methods since the original 1989 version, and provides an updated compilation of the most cost-effective and scientifically valid approaches.

The evaluation of physical, chemical (physiochemical constituents), and biological components are all collected as part of the RBP method to determine water quality of streams.

The Georgia Ecoregions Project conducted the following protocols based on the RBP document: (1) physical characteristics and water quality; (2) visual-based habitat assessment; (3) macroinvertebrate collection–multihabitat approach: D-frame dip net (basic RBP protocol focuses on state's needs and state's current methods at the time); (4) sorting/subsampling (basic principles as required by the RBP and developed to meet the state's needs); (5) quality control/quality assurance; and (6) macroinvertebrate multimetric index. These methods, along with others conducted as part of the Ecoregions Project, are discussed in the following sections. Although the methods described below are specific to a project in Georgia, these protocols can be easily adapted to other ecoregion-based projects and should be viewed as general guidelines for the development of an RBP sampling scheme and can incorporate specifications recommended by cooperating resource agencies.

MATERIAL AND METHODS SPECIFIC
TO THE GEORGIA ECOREGIONS PROJECT

STUDY AREA

The study area for the Ecoregions Project included the entire state of Georgia as well as catchments of the target size (≈ 10 to 100 km^2) shared with the neighboring states of Alabama, Florida, North Carolina, and Tennessee. No catchments of the target size were shared with South Carolina. The study area comprises six ecoregions as described by Omernik (1987): (a) the Blue Ridge Mountains, (b) the Ridge and Valley, (c) the Southwestern Appalachians, (d) the Piedmont, (e) the Southeastern Plains, and (f) the Southern Coastal Plain (Figure 3.1).

Geographic information system (GIS) programs were used to delineate potential reference quality catchments of the target size with minimal anthropogenic influence (Olson 2002); that is, those remaining after applying the following filters:

- At least 80% catchment area within the Level III or IV ecoregion of interest
- Minimal upstream impoundments
- No known National Pollutant Discharge Elimination System (NPDES) discharges upstream
- No known spills or other pollution incidents

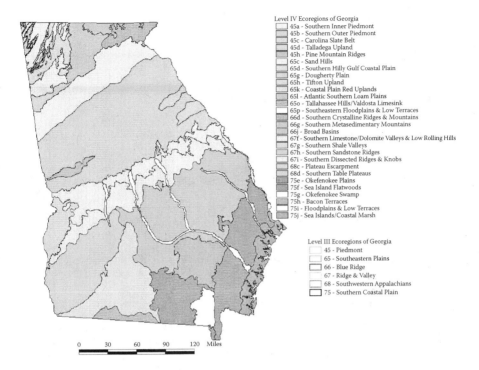

Level IV Ecoregions of Georgia
- 45a - Southern Inner Piedmont
- 45b - Southern Outer Piedmont
- 45c - Carolina Slate Belt
- 45d - Talladega Upland
- 45h - Pine Mountain Ridges
- 65c - Sand Hills
- 65d - Southern Hilly Gulf Coastal Plain
- 65g - Dougherty Plain
- 65h - Tifton Upland
- 65k - Coastal Plain Red Uplands
- 65l - Atlantic Southern Loam Plains
- 65o - Tallahassee Hills/Valdosta Limesink
- 65p - Southeastern Floodplains & Low Terraces
- 66d - Southern Crystalline Ridges & Mountains
- 66g - Southern Metasedimentary Mountains
- 66j - Broad Basins
- 67f - Southern Limestone/Dolomite Valleys & Low Rolling Hills
- 67g - Southern Shale Valleys
- 67h - Southern Sandstone Ridges
- 67i - Southern Dissected Ridges & Knobs
- 68c - Plateau Escarpment
- 68d - Southern Table Plateaus
- 75e - Okefenokee Plains
- 75f - Sea Island Flatwoods
- 75g - Okefenokee Swamp
- 75h - Bacon Terraces
- 75i - Floodplains & Low Terraces
- 75j - Sea Islands/Coastal Marsh

Level III Ecoregions of Georgia
- 45 - Piedmont
- 65 - Southeastern Plains
- 66 - Blue Ridge
- 67 - Ridge & Valley
- 68 - Southwestern Appalachians
- 75 - Southern Coastal Plain

0   30   60   90   120   Miles

FIGURE 3.1   Level III and IV ecoregions of Georgia.

- Low human population density
- Low agricultural activity
- Low urbanization
- Low silvicultural activity
- Low road and highway density
- Minimal nonpoint source (NPS) pollution problems
- No known intensive fish stocking

The combination of catchment-wide land use evaluation and consideration of direct impacts to stream and riparian areas (road crossings, fish stocking, and so forth) provides for a mechanism that selected only those streams that are minimally impaired and representative of ecoregions under consideration (Olson 2002). Land use and land cover data were obtained from the 1998 Land Cover Map of Georgia (Natural Resources Spatial Analysis Laboratory 2001). Catchment-wide land use was quantified using the Arcview extension Analytical Tools Interface for Landscape Assessments (ATtILA) (Ebert and Wade 1999). ATtILA is intended to analyze landscape data, and is well suited for use in quantifying relative human impacts on stream systems from NPS sources. Table 3.1 summarizes primary and secondary criteria used to rank candidate reference streams for minimal anthropogenic influence (adapted from Olson 2002).

TABLE 3.1
Land Use Measures to Select Candidate Reference Streams

Primary Selection Measures		Secondary Selection Measures	
Catchment-Wide	% Urban	45-m Riparian Buffer	% Urban
	% Total Agriculture		
			% Total Agriculture
	% Barren		
			% Barren
	Density Road/Stream Crossings		
	Road Density	135-m Riparian Buffer	% Urban
	Impoundment Density		
15-m Riparian Buffer	% Urban		
			% Total Agriculture
	% Total Agriculture		
			% Barren

Land use and land cover data derived from the 1998 Land Cover Map of Georgia are included in Appendix A. Prioritization of candidate reference stream catchments (by Level III and IV ecoregions) based upon anthropogenic land use was necessary to target field sampling efforts toward those candidate reference streams most representative of the best available reference conditions in given ecoregions.

Sample Collection

Benthic macroinvertebrate and water chemistry samples were collected during an index period to limit seasonal variability associated with such data. In Georgia, for example, during three successive sample seasons (September 1999–February 2000; September 2000–February 2001; and September 2001–February 2002) 119 candidate reference stream samples, 125 impaired stream samples, and 34 randomly selected quality control (QC) duplicate samples were collected in all major ecoregions and subecoregions in Georgia.

Sampling Site Selection

Most sample locations chosen are at road/bridge crossings as far downstream in the target catchment as possible. In some cases, samples are either not collected or samples are taken farther upstream than intended due to lack of access, lack of sufficient flow, or anomalous characteristics of physical and in-stream habitat. When possible, samples are collected at least 200 m upstream of bridge crossings to preclude the influence of hydrologic modification by the bridge or runoff from the roadway.

In a few cases, land cover data proved to be out of date, and actual field conditions in candidate reference catchments were not indicative of the predicted land use. Stream segments identified *a priori* as candidate reference streams (based on remote GIS data) and that exhibited evidence of habitat alteration or NPS pollution were either not sampled, or were sampled and not included in the characterization of the reference stream condition for each ecoregion (not minimally impaired and considered subreference).

At a representative stream segment, a 100 m stream reach is delineated and flagged at 0, 50, and 100 m. Usually the 0 m mark (downstream) is at a hydraulic control point (for example, a riffle). GPS coordinates are taken and noted on all applicable field sheets.

FIELD SAMPLING

All field sampling methods should conform to the EPA's Rapid Bioassessment Protocols and to state field sampling protocols (Barbour et al. 1999; GAEPD 1999). Standard operating procedures (SOPs) and field sheets for all field sampling methods and observations may be found in the "Quality Assurance Project Plan, Georgia Ecoregions Project, Phase II" (Columbus State University 2000). Chain-of-custody is closely monitored and documented from field collection through laboratory and data analysis. Field sampling QC is achieved through the collection and analysis of duplicate samples and field blanks (for chemical samples) at a randomly selected 10% of all stream sites sampled.

Water chemistry grab samples are then collected in flowing water in precleaned and acid-washed 1 L or 500 mL bottles. Grabs are collected at the 0 m mark, and sample locations are approached from downstream. Chemical samples were preserved in the field with 2.0 mL of nitric acid (metals) or sulfuric acid (nutrients). Samples were refrigerated during transportation from the field to the laboratory for analysis of alkalinity, hardness, ammonia, nitrite, nitrate, phosphorus, zinc, manganese, copper, and iron.

In situ environmental measurements were recorded for air and water temperature (°C), dissolved oxygen concentration (mg/L) and percent saturation, pH, conductivity (μs/cm), turbidity (NTU), and water depth. Values are measured with a Hydrolab H20 multiprobe sonde and recorded on applicable field sheets.

Biological sampling is performed using multihabitat benthic macroinvertebrate samples with reallocation of d-frame net jabs (or sample units [SU]) for "missing" habitat types. Separate hierarchies of sample priority are used for high- and low-gradient streams (Table 3.2). Stream gradient is determined based on the presence of typical features associated with a particular stream type (e.g., gravel or cobble riffles in high-gradient streams).

As an example, if a low-gradient stream reach to be sampled was absent woody debris and snags, samples would be reallocated to the remaining habitat types present in order of Table 3.2 (e.g., 9 SU taken from undercut banks/rootwads, 6 from leaf packs, and 5 from sand). A total of at least 20 SU (but never more than 23) were taken at each stream reach regardless of prevalence or quality of habitat types present. Macrophytes were only sampled if present and were not included in the reallocation hierarchy. Samples were collected beginning at the 0 m mark of the reach, moving upstream. Sample units were distributed throughout the entire

TABLE 3.2
Benthic Sample Priority

High-Gradient Streams		
Priority	Habitat Type	Number of SU
1	Fast Riffles	3
2	Slow Riffles	3
3	Snags	5
4	Undercut Banks/Rootwads	3
5	Leaf Packs	3
6	Sand	3
7	Macrophytes (if any)	3
Low-Gradient Streams		
Priority	Habitat Type	Number of SU
1	Woody Debris/Snags	8
2	Undercut Banks/Rootwads	6
3	Leaf Packs	3
4	Sand	3
5	Macrophytes (if any)	3

100 m reach. All material (detritus, minerals, and macroinvertebrates) was combined into a single composite sample and washed, streamside. Large material is removed from the sample prior to preservation, but only after thoroughly having checked for attached or concealed organisms. Each stream sample is packed into labeled 1 L plastic bottles and preserved with 70% ethanol. Benthic field sheets are completed to indicate habitat types sampled and provided a qualitative listing of biota encountered. Samples are transported to the laboratory for processing and taxonomic identification.

Physical measurements and observations at each station include a habitat assessment, physical characterization evaluation, stream cross-sectional profile, and modified Wolman pebble count.

Visual-based physical habitat assessments are conducted on each 100 m stream reach. The habitat assessment protocol is dependent upon whether the stream under consideration is high- or low-gradient. The RBP habitat assessment considers the quality and variety of both in-stream (substrate and channel morphology) and riparian (bank structure and vegetation) habitat features (Barbour et. al. 1999). Stream reaches are evaluated on a 200-point scale with individual components accounting for a maximum of either 10 or 20 points. Scoring is divided into four categories: optimal, suboptimal, fair, and poor.

Physical characterization of streams is also important in evaluating the quality and integrity of stream habitat. The RBP physical characterization form documents such pertinent physical qualities as weather conditions, site location, watershed features and surrounding land use, stream type, riparian vegetation, and in-stream features. These data are helpful supplements to better understand the role of physical habitat in stream water quality and biotic integrity. Physical characterization is conducted following the habitat assessment at each site visited. Characterization observations are also conducted by consensus among trained field personnel present. Measurements and observations are documented and recorded.

Stream cross-sectional profiles are conducted at a representative transect usually close to the 50 m mark of the reach. Iron pins (rebar) are driven into both banks. A measuring tape (tag-line) is stretched taut across the stream perpendicular to the flow of the stream and leveled with a hanging line level. A 2-m stick is used to measure water depth from the tag-line and distance from bank pins. Depth and distance measurements are made in at least 20 intervals across the bank-full width of the stream. More measurements are made in areas of high variability, and fewer measurements are made in areas of homogeneous depth and substrate contour. Velocity estimates are also made using the average of three "runs" of 10 m with a flotation apparatus. Width and depth measurements from the cross-sectional profile, along with velocity estimates, allows for an estimation of stream discharge (using the velocity–area method; Gore 2006). Such baseline data may additionally serve to document changes in channel structure over time.

The modified Wolman pebble count (Wolman 1954) is a method to assess particle size and variability of stream substrates. Pebble counts are conducted at each sample location using 100 particles randomly selected from water-covered portions of the 100 m reach. Sampling effort is spread so that the first particle is picked up near the 0 m mark and the last particle near the 100 m mark. Measurements are made with a sand card (McCollough 1984) for soils (clay, silt, and sands), and small calipers (gravels and small cobbles) or large calipers (cobbles and boulders) for larger inorganic substrates.

LABORATORY ANALYSIS

As with field sampling, laboratory analysis is partitioned by physical, chemical, and biological data. Data entry, database development, and data analysis will be treated separately.

PHYSICAL ANALYSIS

Physical information collected at all stream sites is evaluated for anomalous features. As previously mentioned, a valid reference stream must be representative of the area it seeks to characterize. Realistic goals for impaired streams will not be accurate if based on unrepresentative reference streams (for example, basing expectations only on a single high-gradient, gravel-cobble stream in any ecoregion). Habitat assessment, physical characterization, and Wolman pebble count data are used to refine the candidate reference stream pool (to remove streams from consideration that are not minimally impaired based on local riparian and in-stream habitat). Evidence of human modification (dams, channelization), severely degraded habitat (habitat assessment score greater than 2 standard deviations below the mean for candidate reference streams in an ecoregion), and anomalous substrate (for example, 100% silt in an ecoregion where other candidate reference streams have less than 20% silt and more substrate variability) are physical characteristics used in conjunction with water chemistry parameters to eliminate some streams as candidate reference sites. Stream cross-sectional profiles are used only to estimate discharge and provide additional baseline documentation of channel morphology. Relationships among physical characteristics in Level III and Level IV ecoregions are also documented.

CHEMICAL ANALYSIS

In addition to *in situ* water quality data, samples from all sites are laboratory analyzed for nutrients and metals. Preserved field samples are cold-held for transport to the lab and analyzed within allowable holding times. Several nutrient parameters have very short holding times (less than four days), while the metals analyses allow for holding times up to six months. All chemical laboratory analyses should conform to EPA and state methods and guidelines. Laboratory Quality Control is accomplished by analysis of duplicate samples and laboratory blanks on a randomly selected 10% of all stream samples analyzed. Data are reported and documented on applicable laboratory forms, and unused portions of samples are disposed of in the proper manner.

BIOLOGICAL ANALYSIS

Preserved composite benthic macroinvertebrate samples are held in a locked room in the laboratory until processing. The first step in processing a sample is to prepare the sample for subsampling. A fixed-count random subsample is the preferred technique of the RBPs (Barbour et al. 1999). Subsampling is in accordance with RBP recommendations and follows the protocol set out by Caton (1991). Samples are rinsed in tap water to temporarily remove alcohol residue. The entire sample is then spread evenly on a gridded tray (30 × 36 cm or thirty 6 × 6 cm grid squares) and covered in water to preclude desiccation of the organisms and detritus. Two gridded trays are used in instances where sample material overflows a single tray. Samples are then divided equally between the two trays and spread evenly on each tray. Random number sheets are generated to indicate the 6 × 6 cm grid squares that are selected from the sample and sorted. Each grid square picked is removed and placed in a white picking tray under bright light and sorted with forceps to separate all aquatic organisms from detritus and inorganic material. The target number of organisms for this method is 200 individuals, with 160 to 240 considered within the acceptable range. In no case are less than 4 of the 30 grid-squares picked. If the number of organisms is greater than 240 after four squares, those organisms are respread on a gridded tray and repicked until the target number is achieved. If the total number of organisms is less than 160 after all 30 squares have been picked, the sample is removed from consideration as a candidate reference stream. All benthic macroinvertebrates are preserved in vials of 95% ethanol, labeled, and held for taxonomic identification. All detritus is preserved for quality control purposes. The detritus from a minimum of 10% of all subsamples is rechecked to ensure that laboratory personnel were not missing organisms at an unacceptably high rate.

Taxonomic identification of the subsamples is performed using appropriate taxonomic keys (Appendix B). For the southeastern United States, the primary taxonomic sources are Merritt and Cummins (1996) and Brigham et al. (1982), although a number of keys can be used. Initially, organisms are sorted to order or family level. Taxonomic certainty ratings (TCRs) are assigned to each identification noted. TCRs (on a scale of 1 to 5) serve to indicate how certain the taxonomist is of the identification (with 1 being most certain). In some cases, invertebrates are damaged or missing key body parts that lead to higher TCRs. Identification is to lowest practicable taxonomic level (usually genus) and is recorded on macroinvertebrate bench sheets. At least 10% of all sample identifications by a single trained taxonomist are independently verified with 90% agreement among identifications as the standard. If 90% agreement is not reached, retraining occurrs and reidentification results.

DATA ANALYSIS

Data analysis occurrs in a stepwise process that has been used to develop biocriteria for streams in several states and regions including Maryland (Stribling et al. 1998), Arizona (Gerritsen and Leppo 2000), Idaho (Jessup and Gerritsen 2000), the Mid-Atlantic states (Maxted et al. 2000), West Virginia (Tetra Tech 2000a), Wyoming (Stribling et al. 2000), and Mississippi (Tetra Tech 2002).

Although the intent of the research and the precise methods differ slightly from state to state, the general conceptual framework is the same. After stream site groups are determined and quantitative field data are collected, biological metrics are compiled, calculated, and tested; biotic indices are developed; and indices are tested and refined.

DATABASE DEVELOPMENT AND SITE GROUPS

The first step in the process is the development of a comprehensive database. Following ecoregionalization, site selection and reconnaissance, field sampling, laboratory analysis of chemical parameters and physical measurements and observations, and benthic macroinvertebrate subsampling and taxonomic identification, all data are entered into a database system (the Georgia Ecoregions Project used the Ecological Data Application System [EDAS]), Version 3.3.2 (Tetra Tech 2000b). For example, EDAS is a Microsoft Access© relational database program that allows storage, manipulation, and retrieval of ecological data. Quality control measures for data entry, including the verification of correct entry of a random selection of 10% of all sites evaluated.

Following database development, abiotic factors including *in situ* and laboratory chemical parameters, and physical habitat assessments and characterization are queried from the database and used to refine the candidate reference stream site group to accurately characterize the best attainable biological condition for each ecoregion and subecoregion considered. Site groups for this research are either reference or impaired within the geographic ecoregional framework. Subsequent research has established a gradient of impairment that is used to establish additional site groups such as is used in the RBP habitat assessment (e.g., optimal, suboptimal, fair, poor; Gore et al. 2005). Both reference and impaired (stressor) sites are sampled in all six ecoregions (and thereby 28 subecoregions) under consideration. The refinement of site groups and removal of subreference candidate reference streams (for abiotic considerations) leads to the final group of reference streams that characterize the biological reference condition (Appendix C).

METRIC CALCULATION AND SCORING

Fifty-nine metrics were initially calculated in EDAS, each with a likelihood of being applicable in Georgia and with a documented stress response (Table 3.3). At least 10% of all automated metric calculations were hand-verified to ensure that calculations were correct. In addition, the EDAS queries that perform the calculations were checked to ensure that the formulas were written correctly. Corrections were made when necessary and the database was updated to reflect the changes. Raw data values for all metrics in each Level III and IV ecoregion were exported from EDAS into Microsoft Excel© spreadsheets for ease of manipulation and calculation. A separate spreadsheet was created for each ecoregion. Copies of raw and standardized metric data for all ecoregions of Georgia are available from the primary author.

For each metric, percentile values (5th, 25th, 75th, and 95th), measures of variability (minimum and maximum), and measures of central tendency (mean and median) are calculated for the reference stream distribution. The values from the reference distribution are then used for comparison with metric data from impaired streams sampled and processed in the same manner. Raw data values are evaluated statistically and graphically for their ability to individually discriminate streams in the reference and impaired site groups.

Raw data are evaluated statistically by determining discrimination efficiency (DE). The calculation of DE depends (in part) upon whether an individual metric value is expected to increase or decrease in response to stress. A responsive metric will identify a test stream as being either stressed or nonstressed. Discrimination efficiency is a measure of this ability. Discrimination efficiencies for metrics considered for index development can be found in Appendix D. Raw metric scores for reference and impaired streams in each site class (ecoregion) are used to calculate the DE for each metric with the formula

$$DE = 100 \times a/b,$$

TABLE 3.3
Metric Stress Response

Metric Category	Metric	Stress Response
	Total Taxa	**Decrease[a]**
Richness	Ephemeroptera, Plecoptera, and Trichoptera (EPT) Taxa	Decrease[a]
	Ephemeroptera Taxa	Decrease[a]
	Plecoptera Taxa	Decrease[a]
	Trichoptera Taxa	Decrease[a]
	Coleoptera Taxa	Decrease[b]
	Diptera Taxa	Decrease[a]
	Chironomidae Taxa	Decrease[a]
	Tanytarsini Taxa	Decrease[b]
	Evenness	Decrease[c]
	Margalef's Index	Decrease[c]
	Shannon–Wiener Index	Decrease[b]
	Simpson's Diversity Index	Increase[d]
Composition	% EPT	Decrease[a]
	% Ephemeroptera	Decrease[a]
	% Amphipoda	Decrease[b]
	% Chironomidae	Increase[a]
	% Coleoptera	Decrease[b]
	% Diptera	Increase[a]
	% Gastropoda	Decrease[b]
	% Isopoda	Increase[b]
	% Noninsect	Increase[a]
	% Odonata	Increase[b]
	% Plecoptera	Decrease[a]
	% Tanytarsini	Decrease[a]
	% Oligochaeta	Increase[a]
	% Trichoptera	Decrease[a]
	% Chironominae/Total Chironomidae (TC)	Variable[a]
	% Orthocladiinae /TC	Decrease[a]
	% Tanypodinae/TC	Increase[a]
	% Hydropsychidae/Total Trichoptera	Increase[a]
	% Hydropsychidae/Total EPT	Increase[a]
	% Tanytarsini/TC	Decrease[a]
	% *Cricotopus* sp. and *Chironomus* sp./TC	Increase[a]

(*Continued*)

TABLE 3.3 (CONTINUED)
Metric Stress Response

Metric Category	Metric Total Taxa	Stress Response Decrease[a]
Tolerance/Intolerance	Tolerant Taxa	Increase[a]
	% Tolerant Individuals	Increase[a]
	Intolerant Taxa	Decrease[a]
	% Intolerant Individuals	Decrease[a]
	% Dominant Individuals	Increase[a]
	Dominant Individuals	Increase[a]
	Beck's Index	Decrease[a]
	Hilsenhoff's Biotic Index (HBI)	Increase[a]
	North Carolina Biotic Index (NCBI)	Increase[e]
Functional Feeding Group	% Scraper	Decrease[a]
	Scraper Taxa	Decrease[f]
	% Collector	Decrease[f]
	Collector Taxa	Decrease[f]
	% Predator	Decrease[f]
	Predator Taxa	Decrease[f]
	% Shredder	Decrease[a]
	Shredder Taxa	Decrease[d]
	% Filterer	Increase[a]
	Filterer Taxa	Decrease[a]
Habit	Clinger Taxa	Decrease[a]
	% Clinger	Decrease[a]
	Burrower Taxa	Decrease[a]
	Climber Taxa	Decrease[a]
	Sprawler Taxa	Decrease[a]
	Swimmer Taxa	Decrease[a]

[a] (Barbour et. al. 1999.)
[b] (Barbour et. al. 1996.)
[c] (General literature.)
[d] (Jessup and Stribling 2002.)
[e] (Lenat 1993.)
[f] (Gerritsen and Leppo 2000.)

where a is the number of impaired streams scoring below the 25th percentile of the reference distribution for metrics that decrease in response to stress (or the number of impaired streams scoring above the 75th percentile of the reference distribution for metrics that increase in response to stress), and b is the total number of impaired samples. Values for DE for each metric in each ecoregion (in Georgia) are in Appendix D. Individual metrics in each metric category (richness, composition,

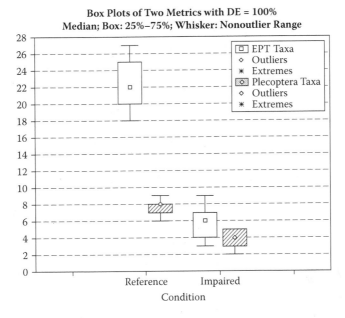

FIGURE 3.2 Box plots for two metrics with discrimination efficiency equal to 100%.

tolerance/intolerance, functional feeding group, and habit) that have the highest DE are evaluated graphically to determine the strength of the discrimination.

Graphical evaluation is essential to truly identify which metrics can differentiate reference from impaired streams as discrimination efficiency only identifies metrics for further study. Metrics may each have a discrimination efficiency equal to 100% while one metric clearly outperforms the other when viewed graphically (Figure 3.2). In this example, EPT taxa is a much stronger metric than Plecoptera taxa because there is a greater difference between the site groups. Box-and-whisker plots of site group distributions (of reference and impaired streams) for metrics with the highest DE in each metric category are evaluated. The best performing metrics exhibit the strongest separation of stressed and nonstressed sites.

Box-and-whisker plots of reference versus impaired metric distributions in each ecoregion are given a score from 0 to 3 (with 3 being the strongest) to indicate the strength of the metric (Barbour et al. 1996). Metrics that show no overlap of interquartile ranges (IQRs) are given a score of 3. Metrics that had some overlap of IQRs but with both medians outside the interquartile overlap are given a score of 2. Metrics with moderate overlap of IQRs but with the median of the impaired sites outside the IQR of the reference sites are given a score of 1. Finally, metrics with extensive overlap of IQR or with both medians in the overlap are given a score of 0 (Figure 3.3).

Metrics that discriminate reference from impaired stream groups based on statistical and graphical analysis (DE and box plot scores) are considered for inclusion in ecoregionally specific macroinvertebrate indices. All metrics that do not differentiate

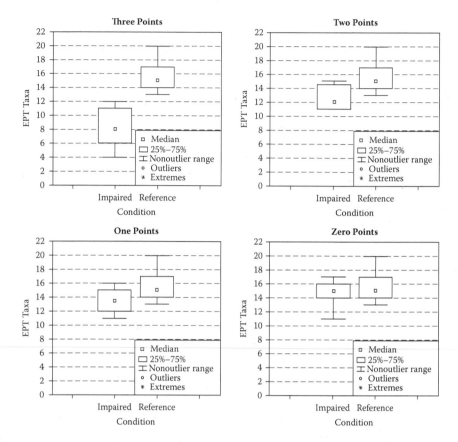

FIGURE 3.3   Box plot scoring system.

reference from impaired streams are removed from consideration for inclusion in the candidate indices. In no case is a metric with a discrimination efficiency less than 50% used. In addition to evaluating the efficacy of the metric, it is also important to check for correlations between metrics to preclude inclusion of redundant metrics in the same index (e.g., using EPT taxa and Plecoptera taxa in the same index when all the EPTs are Plecoptera).

Pearson product-moment correlation analyses (Zar 1999) of raw metric values reveal those that were redundant with one another. When two metrics are calculated as having a Pearson's r-correlation of greater than 0.90 or less than −0.90, the metrics are not considered for use in the same candidate index. When two metrics have a Pearson's r-value of 0.80 to 0.90 or −0.80 to −0.90, the metrics are considered valid for all candidate indices if their relationship is not a linear relationship (as judged by scatterplot). If the scatterplot reveals a linear relationship, the metrics are considered to be codependent and thus are not considered for inclusion in the same candidate index (Tetra Tech 2002).

Metrics with Pearson's r-values between 0.80 and −0.80 are considered for use in all candidate indices. The individual metrics from each metric category that best differentiate reference streams from impaired streams, and meet the aforementioned criteria, are considered for inclusion in candidate indices.

Remaining candidate metrics raw data values are scored, or standardized, for comparison to one another. Many of the metrics are on different scales or have no absolute maximum. In some cases, the number of taxa of a particular family is limited only by the geographic distribution of taxa occurrence. For some metrics, a high raw score indicates higher biological integrity (e.g., EPT taxa), and for some metrics, a low raw score indicates higher biological integrity (e.g., HBI). Standardization allows for all metrics, whether they increase or decrease in response to stress, to be compared on a 100-point scale (with a higher score equaling a "better" score for that metric).

The method of standardization varies depending on whether the metric score increases or decreases in response to stress (Table 3.3). For those that decrease in the presence of stressed conditions, standardized scores are calculated as

$$100 \times c/d,$$

where c is the metric value for the test stream, and d is the 95th percentile value of the reference stream distribution for the site class. And, for those metrics that increase in the presence of stressed conditions, scores are calculated as

$$100 \times [(e - c)/(e - f)],$$

where e is the highest observed value among all streams (reference and impaired) within the site class, c is the metric value for the test stream, and f is the 5th percentile value of the reference stream distribution for the site class.

MULTIMETRIC BIOTIC INDEX DEVELOPMENT

The multimetric approach assimilates biological data, with various ecological meanings, into a single index (for each Level III and Level IV ecoregion evaluated) to gauge the health of a stream. Final indices are often comprised of five to seven metrics with at least one metric chosen from each of the metric categories (richness, composition, tolerance/intolerance, functional feeding group, and habit). In some cases, particular metric categories do not perform well (i.e., all DE values are <50%) in specific ecoregions. The inclusion of such metrics decreases the discriminatory power of the index. In such cases, additional metrics from metric categories that did perform well are used. In a very few instances, in the Georgia Ecoregions Study (particularly among the Level III ecoregions) only 3 or 4 metrics of the initial 59 evaluated had adequate DE, box-and-whisker plot separation, and were minimally redundant. In those cases, the final indices were comprised of the remaining metrics.

Indices are assembled in an additive manner, such that

$$\text{Index Score for Ecoregion (Subecoregion)} = (g + h + i + j + k + l \ldots)/n$$

where $g, h, i, j, k, l, \ldots$ is the standardized score of the best candidate metrics, and n is the total number of metrics included in the index. For the Georgia Ecoregions Study, all final indices contained five to seven metrics, and were scored on a 0- to 100-point scale. Indices are also evaluated for discrimination efficiency and box plots are examined. The candidate index for each Level III and IV ecoregion that provides the highest DE and best graphical spread is selected as the final index.

REFERENCES

Barbour, M.T., J. Gerritsen, G.E. Griffith, R. Frydenborg, E. McCarron, J.S. White, and M.L. Bastian. 1996. A framework for biological criteria for Florida streams using benthic macroinvertebrates. *Journal of the North American Benthological Society* 15: 185–211.

Barbour, M.T., J. Gerritsen, B.D. Snyder, and J.B. Stribling. 1999. *Rapid Bioassessment Protocols for Use in Streams and Wadeable Rivers: Periphyton, Benthic Macroinvertebrates and Fish*, 2nd ed. EPA 841-B-99-002. Washington, DC: U.S. Environmental Protection Agency, Office of Water.

Bolstad, P.V., and W.T. Swank. 1997. Cumulative impacts of land use on water quality in a Southern Appalachian watershed. *Journal of the American Water Resources Association* 33(3): 519–533.

Brigham, A.R., U. Brigham, and A. Gnilka, eds. 1982. *Aquatic Insects and Oligochaetes of North and South Carolina.* Mahomet, IL: Midwest Aquatic Enterprises.

Caton, L.W. 1991. Improved subsampling methods for the EPA "rapid bioassessment" benthic protocols. *Bulletin of the North American Benthological Society* 8(3): 317–319.

Columbus State University. 2000. *Quality Assurance Project Plan, Georgia Ecoregions Project, Phase II.* Columbus, GA: Columbus State University, Department of Environmental Science.

Ebert, D.W., and T.G. Wade. 1999. *Analytical Tools Interface for Landscape Assessments (ATtILA) User Guide* [ArcView GIS extension], version 2.0. Las Vegas, NV: U.S. Environmental Protection Agency, Office of Research and Development, National Exposure Research Laboratory, Environmental Sciences Division, Landscape Ecology Branch.

Georgia Department of Natural Resources (GAEPD). 1999. *Standard Operating Procedures: Freshwater Macroinvertebrate Biological Assessment* (Draft 1999). Atlanta: GAEPD, Environmental Protection Division, Watershed Protection Branch.

Gerritsen, J., and E.W. Leppo. 2000. *Development and Testing of a Biological Index for Warm Water Streams of Arizona.* Prepared for Arizona Department of Environmental Quality. Owings Mills, MD: Tetra Tech, Inc.

Gibson, G.R., M.T. Barbour, J.B. Stribling, J. Gerritsen, and J.R. Karr. 1996. *Biological Criteria: Technical Guidance for Streams and Small Rivers* (rev. ed.). EPA 822-B-96-001. Washington, DC: U.S. Environmental Protection Agency, Office of Science and Technology.

Gore, J.A. 2006. Discharge measurements and stream-flow analysis. In *Methods in Stream Ecology.* (2nd ed.), eds. F. Hauer and G. Lamberti, 51–77. San Diego, CA: Academic Press.

Gore, J.A., J.R. Olson, D.L Hughes, and P.M. Brossett. 2005. *Reference Conditions for Wadeable Streams in Georgia with a Multimetric Index for the Bioassessment and Discrimination of Reference and Impaired Streams.* Atlanta: Georgia Department of Natural Resources.

Jessup, B., and J. Gerritsen. 2000. *Development of a Multimetric Index for Biological Assessment of Idaho Streams Using Benthic Macroinvertebrates.* Boise: Idaho Department of Environmental Quality.

Jessup, B.K., and J.B. Stribling. 2002. *Further Evaluation of the Wyoming Stream Integrity Index, Considering Quantitative and Qualitative Reference Stream Criteria.* Prepared for U.S. Environmental Protection Agency, Region 8. Owings Mills, MD: Tetra Tech, Inc.

Johnson, L.B., C. Richards, G.E. Host, and J.W. Arthur. 1997. Landscape influences on water chemistry in Midwestern stream ecosystems. *Freshwater Biology* 37: 193–208.

Lenat, D.R. 1993. A biotic index for the Southeastern United States: Derivation and list of tolerance values, with criteria for assigning water-quality ratings. *Journal of the North American Benthological Society* 12: 279–290.

Maxted, J.R., M.T. Barbour, J. Gerritsen, V. Poretti, N. Primrose, A. Silvia, D. Penrose, and R. Renfrow. 2000. Assessment framework for mid-Atlantic coastal plain streams using benthic macroinvertebrates. *Journal of the North American Benthological Society* 19(1): 128–144.

McCollough, W.F. 1984. Sand gauge. To order contact: W.F. McCollough, 3100 Elk Ridge Court, Beltsville, MD, 20705.

Merritt, R.W., and K.W. Cummins. 1996. *An Introduction to the Aquatic Insects of North America,* 3rd ed. Dubuque, IA: Kendall/Hunt.

Natural Resources Spatial Analysis Laboratory. 2001. *1998 Land Cover Map of Georgia.* Institute of Ecology, Athens, GA. http://narsal.ecology.uga.edu/

Olson, J.R. 2002. Using GIS and data to select candidate reference sites for stream bioassessment. Master's thesis, Columbus State University, Georgia.

Omernik, J.M. 1987. Map supplement: Ecoregions of the conterminous United States. *Annals of the Association of American Geographers* 77: 118–125.

Stribling, J.B., B.K. Jessup, J.S. White, D. Boward, and M. Hurd. 1998. *Development of a Benthic Index of Biotic Integrity for Maryland Streams.* CBWP-EA-98-3. Annapolis: Maryland Department of Natural Resources, Monitoring and Non-Tidal Assessment Division.

Stribling, J.B., B.K. Jessup, and J. Gerritsen. Tetra Tech, Inc. 2000. Development of biological and physical habitat criteria for Wyoming streams and their use in the TMDL process. U.S. Environmental Protection Agency, Region 8.

Tetra Tech, Inc. 2000a. *A Stream Condition Index for West Virginia Wadeable Streams.* Prepared for U.S. Environmental Protection Agency, Region 3. Owings Mills, MD: Tetra Tech, Inc.

Tetra Tech, Inc. 2000b. *Ecological Data Application System (EDAS): A User's Manual.* Prepared for U.S. Environmental Protection Agency, Office of Water and Oceans and Wetlands, Washington, DC. Owings Mills, MD: Tetra Tech, Inc.

Tetra Tech, Inc. 2002. *Development and Application of the Mississippi Benthic Index of Stream Quality.* Prepared for Mississippi Department of Environmental Quality, Water Quality Assessment Branch. Owings Mills, MD: Tetra Tech, Inc.

Wolman, M.G. 1954. A method of sampling materials in gravel bed streams. *Transactions of the American Geology Union* 35: 951–956.

Zar, J.H. 1999. *Biostatistical Analysis.* (4th ed.) Upper Saddle River, NJ: Prentice Hall.

4 Candidate Reference Conditions

John R. Olson, Duncan L. Hughes,
James A. Gore, and Michele P. Brossett

CONTENTS

BACKGROUND OF BIOASSESSMENT AND THE REFERENCE CONDITION

Our nation's need to improve the condition of streams is mandated by Congress in the Federal Water Pollution Control Act and as modified as the Clean Water Act (CWA, 33 U.S.C. Section *et seq.*). The objective of both acts is "to restore and maintain the chemical, physical and biological integrity of the nation's waters." (CWA, Section 101(a), 33 U.S.C., 1251(a), 1999). To meet these objectives, the law requires states to develop and enforce water quality criteria based on biological assessment (bioassessment) in accordance with methods published by the Environmental Protection Agency (EPA), whenever numerical criteria for toxic pollutants are unavailable (CWA, Section 303(c) (2)(B), 33 U.S.C., Section 1313(c)(2)(B), 1999). The law also requires streams to be assessed for nonpoint source pollution (NPS; CWA, Section 319, 33 U.S.C., Section 1329, 1999), and that the current quality of surface waters and the extent that they will support wildlife be reported to the EPA (CWA, Section 305(b)(1)(B), 33 U.S.C., Section 1313(c)(2)(B), 1999). To support these requirements the EPA has published guidelines on developing bioassessments, the *Rapid Bioassessment Protocols for Use in Streams and Wadeable Rivers: Periphyton, Benthic Macroinvertebrates, and Fish* (Barbour et al. 1999); and biocriteria, *Biological Criteria: Technical Guidance for Streams and Small Rivers* (Gibson et al. 1996). Bioassessments, and their resulting biocriteria, are effective methods for assessing water quality because they provide a way of integrating the chemical, physical, and biological effects on water quality by directly measuring biologic integrity and indirectly measuring physical and chemical integrity. This indirect measurement of chemical integrity may actually be more representative of the true integrity of the stream than base flow chemical data since chemical composition varies widely over time, both seasonally and during storm flows (Bolstad and Swank 1997; Johnson et al. 1997).

The Rapid Bioassessment Protocol (RBP) is a method of gathering, analyzing, and reporting biological, chemical, and physical habitat data to support a number of management decisions to improve water quality. The basic output of these assessments is a characterization of the amount of impairment on an aquatic system (Fore et al. 1996). This information then would support management decisions such as evaluating nonpoint source pollution and the total maximum daily loads (TMDLs) and best management practices (as defined in CWA, Section 303(d)) to control nonpoint sources of pollution. It would also identify areas and standards for restoration, isolate causes and sources of impairment, monitor trends, provide risk management, and help set priorities for restoration (Barbour et al. 1999).

The first step in developing a rapid bioassessment protocol for a specific area is to establish a control, or some desired state of ecological integrity, which is usually the best available condition, referred to as the *reference condition* (Resh 1995). The reference condition is defined as "the condition that is representative of a group of minimally disturbed sites organized by physical, chemical and biological characteristics" (Reynoldson et al. 1997) and serves as a control for all other comparable streams. The reference condition is then characterized by measuring its chemical, physical, and biologic attributes. These same variables can then be measured at known impaired sites in the same geographic region to determine which biological measurements best detect the difference between reference and impaired conditions. These measurements (or metrics) can then be combined into a single number describing the amount of damage to any system that shares the same potential natural condition (i.e., within the same ecoregion). Richards et al. (1997) cautioned that metrics of this sort must be calibrated to each ecoregion if they are to detect changes due to anthropogenic impact from natural variability. Ecoregions are "regions of relative homogeneity in ecological systems or in relationships between organisms and their environment" (Omernik 1987, p. 123).

CHARACTERIZING REFERENCE CONDITIONS: SELECTION OF REGIONAL REFERENCE SITES

Reynoldson et al. (1997) summarized the seven primary methods used to characterize the reference condition (as developed by Hughes 1995 and Johnson et al. 1993) (see Table 4.1). Reynoldson et al. (1997) and Hughes (1995) concluded that the primary method for determining the reference condition should be regional reference sites; that is, a group of sites that will characterize the best attainable condition for a given region (Reynoldson et al. 1997).

TABLE 4.1

Summary of Approaches for Determining Reference Conditions of Streams

Approach	Application	Limitations
Regional Reference Sites	Ordination and indicator analyses are used to determine representativeness of reference sites; acceptable levels of disturbance must be determined	Aquatic ecoregions are applicable to whole faunal assembly but there is some difficulty in applying ecoregions to particular communities; habitat classifications are still needed
Historical Data	Useful if sites have been periodically resampled or if making general statements about conditions	Usually limited to a single invertebrate community; often comparisons with historical data only can reflect serious deterioration; data incomplete or method sometimes unknown
Paleoecological Data	Essentially limited to lakes, diatoms, and chironomids	Poorly suited to streams; diatoms, the most widely used group, are affected by changes in water quality but perhaps less from changes in habitat structure or introduced species
Biotic Indices	For comparison with predetermined hierarchy of values	Conditions represented by indices may not be obtainable because of habitat differences; tolerance usually developed for organic contamination
Experimental Laboratory Data	Relationships between test and species and some stressors (specific toxins, temperatures, etc.) are well known so field data may be used to exclude some reference sites	Data not applicable to entire invertebrate communities and are unsuitable for systems disturbed by other stressors
Quantitative Methods	Plotting metric values against disturbance or natural variables can establish reference conditions through curve fitting	Outliers, uneven distribution of data, and absence of data from minimally disturbed sites can distort models
Best Professional Judgment	Convening expert panels to determine reference conditions and peer review of data and conclusions	Value of judgment is a function of scientists' expertise and the quality of data supplied to them

Source: Adapted from Reynoldson et al. (1997).

To determine the appropriate region for comparisons, each area of interest is initially divided into ecoregions and then subecoregions, which are delineated by grouping areas with similar climate, physiography, geology, soils, and vegetation (Hughes et al. 1986; Omernik 1987). These subecoregions are initially tested to see if they adequately group streams by common or distinctive physical and chemical characteristics. The sites least impaired by human activity are then chosen within each subecoregion; then, these candidate streams are sampled and characterized to establish the reference condition. The reference condition for an ecoregion or subecoregion is an amalgam of conditions typical of the region and does not represent any one stream.

The certainty that a stream is the least impaired is only possible if the condition of all streams is known or is estimated based on probabilistic sampling. This knowledge of stream condition has been used to accurately define the reference condition, resulting in sensitive metrics for determining whether a stream is impacted. An analysis of the streams of the Umpqua National Forest, and Roseburg and Medford Districts of the Bureau of Land Management by Fore et al. (1996) used data on logging, stream condition, and taxonomic richness to choose the 7 least impacted streams from a set of 80 streams. The Maryland Biological Stream Survey, using probabilistic sampling design to collect data from over 1000 streams, established a set of criteria that described the best available conditions found within the state. These criteria (Table 4.2) were used to describe an adequate number of minimally impaired sites (Stribling et al. 1998).

Although results of these previous comprehensive studies are encouraging, we believe that a more practical method, in regions where less data are available, is to identify those streams with the greatest likelihood of having the least impairment and then to measure biological and physiochemical parameters to determine their actual condition. These streams are candidate reference sites. The difficulty in determining which sites reflect reference conditions was summed up by Resh (1995, p. 112), "If reference sites could be distinguished easily from test sites, impairment could be assessed without any measurement." The ability of science to define reference sites

TABLE 4.2
Maryland Reference Site Criteria

pH > 6 (if blackwater stream, ph < 6 and DOC < 8 mg/l)

Acid neutralization capability (ANC) > 50 meq/l

Dissolved O_2 < 4 ppm

Nitrate-N < 4.2 mg/l

Urban land use < 20% of catchment area

Forested land use > 25% of catchment area

In-stream habitat optimal or suboptimal

Riparian buffer width > 15 m

No channelization

No point source discharges

Remoteness rating optimal or suboptimal

Aesthetics rating optimal or suboptimal

was identified as one of the three major limitations of the current system of establishing biocriteria (Science Advisory Board [SAB] 1993).

Hughes (1995) suggested an iterative process to identify reference conditions composed of the following steps:

1. Define areas of interest, using natural boundaries whenever possible.
2. Define water bodies of interest by type (lake, stream, wetland) and size.
3. Delineate (select) candidate reference catchments, focusing on rejecting impacted sites while retaining minimally disturbed sites. This step is done by evaluating available data/maps, by remote sensing, and by obtaining input on candidate sites from local experts and managers.
4. Conduct aerial and field reconnaissance to choose sites and confirm selections. Hughes recommends the use of a qualitative habitat evaluation form.
5. Subjectively evaluate quality of candidate reference sites.
6. Determine the number of sites needed.
7. Quantitatively evaluate reference sites.

Hughes also urged increased objectivity in determining the reference condition, but recognized that the process still relies heavily upon best professional judgment (BPJ) (in terms of available time and budgets). However, regardless of these recommendations, reference sites often have been more subjectively chosen and often are more impaired than is desirable (Hughes et al. 1986; Hughes 1995). For example, the state of Montana limited its search for reference sites to only sites where fish data were already available, resulting in selecting only 38 streams as reference sites for the entire state (Gibson et al. 1996).

Hughes et al. (1986) recommended that the location of candidate sites be discussed with knowledgeable resource managers and scientists, questioning them about the conditions of local streams and about the location of other candidate sites. This recommendation, in absence of a viable alternative, has been used extensively for selecting candidate reference sites. BPJ relies on the knowledge of local experts or on data gathered in previous studies that indicates where the best available aquatic communities are located. In many cases, the resource agency trying to establish the reference conditions already has extensive data collected from attempting previous studies or has its own experience that indicates where potential reference sites are located (Stribling et al. 1998). As an example, the Texas Aquatic Ecoregions Project performed by the Texas Natural Resource and Conservation Commission (TNRCC) analyzed maps of land surface form, potential vegetation, soil type, and land use to choose the catchments that lacked urban development, channelization, and with streams and rivers that flowed through catchments with natural vegetation for that region (Hornig et al. 1995). Bailey et al. (1998) used streams without any point sources of pollution (mining activity) as reference sites, since that was the predominate type of anthropogenic impact in the central Yukon Territory.

Reference sites chosen by BPJ are often limited in both number of stream sites chosen and quality of the site characteristics because of the limited amount of previous research and the fact that BPJ is limited only to streams that are generally easily accessed (Hughes 1995). Since ease of access (road/steam crossings) is a form of anthropogenic disturbance, sites chosen by BPJ may not always be the least impaired streams available. This suggests that a relatively comprehensive yet unbiased analysis of physical, chemical, and biological features in each geographic region might provide a more complete suite of candidate reference sites.

ASSESSING ANTHROPOGENIC IMPACT ON STREAM QUALITY USING THE SYNOPTIC APPROACH

The synoptic approach for determining the amount of cumulative impact of multiple anthropogenic stressors was originally developed as a framework for comparing landscape units that quickly determined the relative amount of anthropogenic impact on a wetland (Abbruzzese and Leibowitz 1997). This approach is a compromise between the need for rigorous results and the need for timely information and is appropriate when little quantitative information is available, the cost of improving these data is high, there is an urgent need to make decisions, and the cost of a wrong answer is low. Bolstad and Swank (1997) demonstrated that this approach could also be used to assess the cumulative impact of NPS pollution on water quality in streams, showing a consistent and cumulative decrease in water quality with increasing nonforest land use, principally building and road density and agricultural land use.

The synoptic approach to assessing anthropogenic disturbance is based upon the idea that disturbance can be estimated by examining the land use in the catchment feeding the stream, influencing or controlling stream conditions (Richards and Host 1994). It follows, that the amount of disturbance in the catchment should predict the extent or intensity of impairment in the stream. Anthropogenic land use affects stream communities directly through changes in water chemistry by affecting the amount of metals and nutrients (Bolstad and Swank 1997), as well as through suspended sediment loading (Lenat and Crawford 1994; Johnson et al. 1997). A clear, negative correlation between the amount of urbanization in a catchment and a stream's biological integrity has been shown in several studies (Lenat and Crawford 1994; Wang et al. 1997; Kennen 1999; Roth et al. 1999), whereas a positive correlation has been shown with the proportion of the catchment that is forested (Roth et al. 1996; Wang et al. 1997; Rothrock et al. 1998; Kennen 1999; Roth et al. 1999).

The correlation between agricultural land use and stream integrity is much less consistent. Although some studies have shown that agriculturally dominated catchments have impaired biological integrity (Richards et al. 1996; Roth et al. 1996; Rothrock et al. 1998), Kennen's (1999) study did not. Lenat and Crawford (1994) found changes related to the amount of agricultural land use in benthic macroinvertebrates but not in fish communities. Roth et al. (1999) found a positive correlation between the fish Index of Biotic Integrity (IBI) and Hilsenhoff's Biotic Index and the amount of agriculture, but no correlation between their benthic IBI and the amount of agriculture. Lammert and Allen (1999) found a weak correlation between the amount of agriculture and fish IBI, benthic IBI, and four other common metrics, but only for land use within 100 m of the stream. Wang et al. (1997) found the impact on fish communities to be nonlinear, with the effects of agriculture only becoming apparent in catchments with more than 50% agriculture. Rothrock et al. (1998) also showed that both increasing road density and silviculture lead to lower biologic integrity. Schnackenberg and MacDonald (1998) also found a strong correlation between the number of road crossings and the percentage of fine particles in the substrate that would affect aquatic communities. They found a weaker correlation between fine particles and the amount of clear-cut forests.

Even though the relationship of stream condition to catchment condition seems clear enough, several factors make the relationship complicated. Several studies have demonstrated that other factors have equal or greater impact than land use patterns on aquatic communities, including geology, topography, and geographical characteristics (catchment area, altitude, and length; (Richards et al. 1996; Johnson et al. 1997; Bailey et al. 1998). There are also interactions between these geologic or geographic features and land use that are difficult to separate. Most agree that catchment land use has more impact on stream communities than the land use of the riparian buffer (Roth et al. 1996; Wang et al. 1997; Kennen 1999). However, Richards et al. (1997) found reach scale properties to be more predictive of species traits than catchment properties, although the catchments may have had an indirect effect on the reach scale properties. Lammert and Allen (1999) also found more of the variance in stream communities to be explained by the type of land use within a 100 m riparian buffer than in the entire catchment. Lammert and Allen proposed that larger scale, less spatially expansive investigations are more sensitive to local changes in physical habitat than smaller scale investigations. The stronger relationship of catchment characteristics to the stream community structure found by others may also have been a function of the precision of the data used. For example, coarse-grained data (larger than 2 ha grain size) were only able to examine the effect of 100 m buffers; a finer grain data would allow for the analysis of smaller and perhaps more influential buffer zones.

If the amount of anthropogenic land use within a catchment is going to be used to predict the relative amount of stream impairment, then these geologic or geographic factors that also effect stream communities must be controlled during analysis. Using an *a priori* classification, by subecoregion, and examining stream catchments within a single order of magnitude, variability in these geologic or geographic factors will be taken into account. Combining catchment-wide land use with measurements of direct impact on streams caused by road crossings and alterations within the riparian zone of streams, such as roads and agricultural land use, should create a measure of the extent of impairment to a stream's ecosystems relative to other systems in the same subecoregion. Schnackenberg and MacDonald (1998) provided an example of this approach by defining their reference sites in terms of cumulative land use and direct impact. They arbitrarily selected reference streams as those with less than 2% of the catchment clear-cut and lowest density of road crossings and roads within 60 m of streams of less than $0.25/km^2$ catchment size.

Our study of the state of Georgia included any catchments shared with the neighboring states of Tennessee, Alabama, North Carolina, and Florida, covering an area of 153,169 km^2. South Carolina did not share any catchments of the size evaluated in this study. This area lies across five ecoregions as described by Omernik (1987): (a) the Blue Ridge Mountains, (b) the Ridge and Valley, (c) the Southwestern Appalachians, (d) the Piedmont, (e) the Southeastern Plains, and (f) the Coastal Plains. These ecoregions categorize the major differences found in topography, physiography, climate, elevation, hydrology, vegetation, wildlife, land use, and surface geology as reflected by soils across Georgia. Each of these ecoregions was further divided into more homogeneous subecoregions, reflecting higher resolution changes

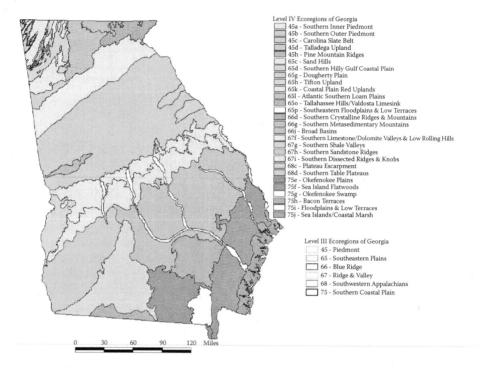

Level IV Ecoregions of Georgia
- 45a - Southern Inner Piedmont
- 45b - Southern Outer Piedmont
- 45c - Carolina Slate Belt
- 45d - Talladega Upland
- 45h - Pine Mountain Ridges
- 65c - Sand Hills
- 65d - Southern Hilly Gulf Coastal Plain
- 65g - Dougherty Plain
- 65h - Tifton Upland
- 65k - Coastal Plain Red Uplands
- 65l - Atlantic Southern Loam Plains
- 65o - Tallahassee Hills/Valdosta Limesink
- 65p - Southeastern Floodplains & Low Terraces
- 66d - Southern Crystalline Ridges & Mountains
- 66g - Southern Metasedimentary Mountains
- 66j - Broad Basins
- 67f - Southern Limestone/Dolomite Valleys & Low Rolling Hills
- 67g - Southern Shale Valleys
- 67h - Southern Sandstone Ridges
- 67i - Southern Dissected Ridges & Knobs
- 68c - Plateau Escarpment
- 68d - Southern Table Plateaus
- 75e - Okefenokee Plains
- 75f - Sea Island Flatwoods
- 75g - Okefenokee Swamp
- 75h - Bacon Terraces
- 75i - Floodplains & Low Terraces
- 75j - Sea Islands/Coastal Marsh

Level III Ecoregions of Georgia
- 45 - Piedmont
- 65 - Southeastern Plains
- 66 - Blue Ridge
- 67 - Ridge & Valley
- 68 - Southwestern Appalachians
- 75 - Southern Coastal Plain

0   30   60   90   120   Miles

FIGURE 4.1   Level III and Level IV ecoregions of Georgia.

in these variables. The subecoregions divide the state into 28 areas, ranging in size from 290 to 31,590 km^2 (see Figure 4.1).

DATA SOURCES AND PREPROCESSING

To conduct a statewide analysis of all wadeable streams, data must be acquired that are spatially expansive, inexpensive, and relatively detailed. Since the amounts of each type of stream impairment that occur may be unknown, the data should be chosen to cover the widest array of potential impairments feasible. This includes agricultural, silvicultural, and urban impacts, as well as road density and road crossings. If possible, data on riparian conditions, point sources of pollution, and hydrologic impacts like in-stream impoundments should be included.

Although the highest possible resolution will yield the best correlations between land use and degree of impairment, a trade-off in data accuracy must always be considered. If data for a catchment is only going to be used relative to other catchments in similar environments, absolute accuracy may not be necessary in some circumstances. If the inaccuracies that occur are randomly or uniformly distributed, then relative measurements of land use between catchments would be unaffected even if the absolute measurements of land use are not completely accurate.

Recently, the competing demands for data that were both high resolution and had statewide coverage would have been insolvable. However, recent gains in desktop computing power, combined with a powerful geographic information system (GIS) capable of spatial analysis, and the ready availability of free spatial data via the Internet has made these types of analyses possible. For example, for analysis of the state of Georgia, a great deal of the basic statewide data described was available at the Georgia GIS Data Clearinghouse, a Web site operated on a contract basis through the Board of Regents of the University System of Georgia with nodes at the University of Georgia and the Georgia Institute of Technology. It provided most common data by county, for the entire state, free of charge to registered users. Data for areas in adjoining states were downloaded either directly from the U.S. Geological Survey (USGS) or from the GIS Data Depot, a licensed distributor of USGS data. A brief description of each of the data sets follows.

Digital elevation model (DEM) data are used to both analyze stream order and to delineate catchment boundaries. DEMs are raster data produced by the USGS that portray surface elevations using a 30 m grid. Each cell of the grid contains its mean elevation. The DEMs are produced by the USGS in standard 7.5 minute quadrangles with a Universal Transverse Mercator (UTM) projection. Using the grid merge script in ArcView, each of these coverages can then be joined with other coverages within the same UTM zone. Where more than one zone is covered, the DEMs of these counties are converted into the predominate projection for that drainage for which that county was a part. This was accomplished using the Spatial Tools 3.3 command for Grid Warp (Hooge 1998). Each of these converted DEMs is then joined to the original DEM set in the other UTM projection to produce two DEM sets (one for each UTM projection) that are divided along hydrologic as opposed to geographic boundaries. DEMs are not entirely error free, with many individual cells missing values. The cells of missing values (no data cells) or cells that have a lower value than surrounding cells will act as infinite sinks when the DEMs are used for hydrologic modeling, such as determining stream order and delineating catchment boundaries. While some of these sinks represent naturally occurring or man-made depressions, most of them tend to be artifacts of the DEMs. These artifacts occur when narrow gullies exist in the drainage areas, so that while the stream is flowing downhill through the gully, the DEM cell with the gully has a higher mean elevation than the upstream cell, giving the appearance of water flowing uphill. Another source of these artifacts occurs when two DEM quadrangles are joined and the downstream quadrangle, while internally consistent, has slightly higher elevations than the upstream quadrangle, producing a set of "dams" across all of the streams draining from one quadrangle to the next. The hydrologic modeling extension to ArcView has a script available for finding and filling any sinks that occur in a DEM by increasing the elevation of the cells in the sink. It will also be necessary to modify this script to find any no data cells, and then replace the missing elevation with the mean of the elevation of all the neighboring cells.

Hydrographic data representing streams, canals, lakes, and reservoirs were used to delineate catchments for analysis and comparison as well as to measure the amount of anthropogenic hydrologic impact within each catchment. Land cover data provided most of the information that was used to determine the amount of disturbance found in both catchments and riparian zones. The most readily available high-resolution data are published as the National Land Cover Data (NLCD) (see, for example, Vogelmann et al. 2001). Another measure of the amount of human impact on a catchment is road density in the catchment. Such data are often available as transportation 1:24,000 scale digital line graphs (DLGs) produced by the USGS or comparable data sets produced by state departments of transportation. Since catchment-wide land use is not necessarily correlated with the occurrence of point sources of pollution, data on point source locations are also needed to ensure streams with point source impacts are excluded as possible reference sites. The EPA has produced a nationwide set of data and a set of GIS analysis tools designed to be used with these data in the Better Assessment Science Integrating Point and Nonpoint Sources (BASINS) program. The data available for point sources included the EPA's Office of Water (EPA/OW) Industrial Facilities Discharge (IFD) database for CONUS (EPA 1998a); the EPA/OW Permit Compliance System (PCS) for CONUS (a national computerized management information system that automates entry, updating, and retrieval of National Pollutant Discharge Elimination System [NPDES] data; EPA 1998b); and the EPA's Toxic Release Inventory (TRI) of industrial manufacturing facilities in the United States.

DATA ANALYSIS

The steps used to choose candidate reference sites are: (1) delineating areas to be compared, (2) measuring the amount of land use impact in each of those areas, and then (3) choosing the least impaired sites to be candidates.

Because stream biota varies longitudinally, any system of comparing streams will have to account for this natural variation (Allen 1997; Stanford 2006). By studying a single size class of stream catchments, this variation is minimized. We have found that in choosing stream size as a focus a balance was struck between using catchments that were small enough to fit within subecoregions and wadeable streams and catchments that were large enough to have perennial streams that would be flowing, even during drought conditions. Fourth-order streams (Strahler 1952) seem to be an appropriate size to study. Fourth-order streams are relatively common even in the smallest subecoregions and are usually still flowing except during severe drought. To increase the number of candidates in a region, it may be useful to add large second- and third-order streams with a total catchment length of more than 8 km and small fifth-order streams with catchment lengths of less than 8 km, all having roughly the same catchment area as most fourth-order streams. As an alternative, Hughes et al. (1986) recommended selecting candidate catchments based on catchment area and annual discharge as opposed to stream order. They also recommended that the sites to be compared differ by less than an order of magnitude.

The dilemma arises about the appropriate status for catchments that lay in multiple subecoregions since subecoregion boundaries do not generally coincide with catchment boundaries. The simplest solution for those streams that reached the required size before crossing the ecoregion boundary is to delineate the catchment at or near that boundary. Streams that are divided into different regions longitudinally or have their headwaters in one subecoregion but do not become large enough before crossing subecoregion boundaries present a more complex dilemma. Different solutions to this dilemma have been proposed. Gibson et al. (1996) recommended that candidate sites lie entirely within a single ecoregion. Warry and Hanau (1993) used sites with at least 50% of the catchment within a single ecoregion. Omernik (1995) recommended using sets of catchments with similar proportions of different ecoregions. Since subecoregion boundaries do not usually represent a demarcation between one area and another, as much as the center of an ecotonal zone between areas (Bryce and Clarke 1996), rejecting streams that did not lie entirely within single subecoregion is likely too harsh a criterion. Any catchment near the boundary will have some of the characteristics of the neighboring region (Gallant et al. 1989). A catchment should be delineated so that it has more characteristics in common with the catchments of its subecoregion than of its neighboring region. We suggest that a decision be made not to delineate those catchments that do not have at least approximately 80% of the catchment within a single subecoregion. This is an arbitrary criterion for delineation, but one that retains catchments representing the entire span of variation found in each subecoregion.

The next step is to measure the relative amounts of human impact upon each catchment. This analysis can be accomplished using an extension for ArcView, called the Analytical Tools Interface for Landscape Assessments (ATtILA), developed by EPA's Landscape Ecology Branch (Ebert and Wade 1999). The "landscape characteristics" function determines the total area and percentage of cover for each of the major land use categories within each catchment. The "riparian characteristics" function is used to calculate total areas and percentages of land use for three buffer zones of different width along all the streams within a catchment. Buffers were calculated, providing buffers with widths of 10 to 15 m, 40 to 45 m, and 130 to 135 m. The "human stresses" function was used to measure road density and the number of stream/road crossings within a catchment. This function is also used to calculate stream density. Although this does not function as a measure of impact, it is used to ensure that each catchment is representative of its subecoregion.

Some measure of disruption of hydrology in each catchment is desired. One of the more effective methods is the measurement of impoundment density because of the availability of data on impounds and the amount of hydrologic impact they create relative to other human disturbances like channelization. The number of surface water withdrawals would also make a good measurement of disturbance. The types of measurements for all catchments are summarized in Table 4.3.

The next step is to decide which catchments are the least impaired, based upon the measurements made. An iterative approach can be used to develop a selection method, starting with comparing the results of different ranking methods against the raw data, then by comparing predictions of stream quality against a sample of ground-truthed streams, and finally to a validation of the method in several different subecoregions. Two basic approaches to selecting candidates from all possible sites in a state would be to either apply a set of criteria as a filter, or develop a ranking system and simply choose the least impaired sites as candidates. Filtering, although simple, does not account for changes in conditions between ecoregions, and will tend to overselect in undeveloped areas and underselect in developed ones. Scoring systems can be used to create lists of the least sites catchments within ecoregions or subecoregions. Sites can be scored simply by determining their quartile for each separate impairment measure. Sites with impairment measurements in the lowest quartile would receive a value of 1, while those in the highest quartile would receive a 4. These scores would then be summed for all measurements and the lowest scoring catchments would be selected as candidate reference sites. Table 4.4 depicts the list of measurements used in the state

TABLE 4.3
Summary of Land Use Measurements

Measurement	Catchment-Wide	15-m Riparian Buffer	45-m Riparian Buffer	135-m Riparian Buffer	Made By	Source Data
% Natural Cover	X	X	X	X	ATtILA	NLCD
% Urban	X	X	X	X	ATtILA	NLCD
% Total Agriculture	X	X	X	X	ATtILA	NLCD
% Row Crops	X	X	X	X	ATtILA	NLCD
% Pasture	X	X	X	X	ATtILA	NLCD
% Barren	X	X	X	X	ATtILA	NLCD
% Forest	X	X	X	X	ATtILA	NLCD
% Wetland	X	X	X	X	ATtILA	NLCD
Road Density	X				ATtILA	USGS/DOT
Density of Road/ Stream Crossings	X				ATtILA	USGS/DOT
Number of Roads/ Stream Crossings	X				ATtILA	USGS/DOT
Impoundment Density	X				Manual analysis	USGS/DOT
Number of Impoundments	X				Manual analysis	USGS/DOT

TABLE 4.4
Land Use Measures Used in Selecting Candidate Reference Sites

Primary Selection Measures

Catchment-Wide	% Urban
	% Total Agriculture
	% Barren
	Road Density
	Density of Road/Stream Crossings
	Impoundment Density
15-m Riparian Buffer	% Urban
	% Total Agriculture
	% Barren

Tie-Breaking Selection Measures

45-m Riparian Buffer	% Urban
	% Total Agriculture
	% Barren
135-m Riparian Buffer	% Urban
	% Total Agriculture
	% Barren

of Georgia. This is not an exhaustive list and other variables can be considered. However, care should be taken so that redundant measures (row crop, pasture, and total agriculture) are eliminated and catchment-wide measures and riparian measures are equally represented. The distribution of the candidate reference sites should adequately cover the described ecoregion or subecoregion. Figure 4.2a shows an example of the candidate reference sites chosen for the state of Georgia and Figure 4.2b shows how the candidate reference catchments were selected (GIS, Best Professional Judgment, and Other).

ASSESSMENT OF REFERENCE SITES

To assess the validity of reference sites, it will be necessary to "ground-truth" predictions of environmental quality by measuring physical, chemical, and biological conditions at the candidate sites. The analysis of these data will allow a more refined choice of the final reference streams. The RBP suggests a suite of physical, chemical, and biological measurements to examine each of the candidate sites (Barbour et al. 1999). This set of measurements should be taken at each of the candidate sites:

1. Benthic macroinvertebrates, collected according to prescribed and consistent field methods, with the macroinvertebrates being identified to the lowest practicable taxonomic level.
2. Water chemistry (Table 4.5), both *in situ* and employing water grab samples that are later tested in the lab.
3. The stream's physical properties, including a stream cross-section, velocity, substrate size using a modified Wolman pebble count, and observations of degree of shading and presence of oils, impacting land uses, bank erosion, and types of deposits.
4. Habitat assessments utilizing RBP habitat assessment methods and forms.

The determination of whether the water quality of the candidate reference sites is impaired is made by comparing the water chemistry data against national standards (for those parameters where they exist) and by comparing them to published data on water quality in other streams in the region. Habitat assessments can also be used to compare candidate sites with existing habitat standards or results of other regional studies. Where published standards are not available, candidate reference sites can be assessed by comparing the 95% confidence interval for each measurement (i.e., habitat, chemical, and so forth) with confidence intervals from known impaired sites to ensure that they do not overlap. Figures 4.3, 4.4, and 4.5 demonstrate typical ranges of information accumulated for candidate reference sites.

To compare the ecological integrity of candidate sites using benthic macroinvertebrate species data, we suggest analysis following the method described in

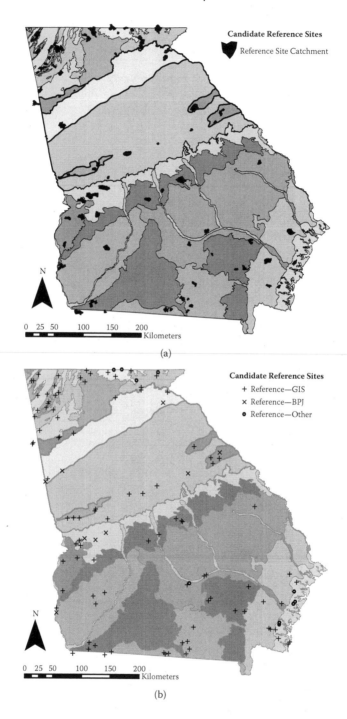

FIGURE 4.2 Candidate sample locations.

TABLE 4.5
Water Chemistry/Quality Parameters Measured at Sites

Parameter Measured	Type of Sample Taken	Method/ Instrumentation Used	Range of Detection
Ammonia (mg/l as N)	Grab Sample	EPA Method #350.3	0.03 to 1400 NH3-N/L
Nitrate (as N)	Grab Sample	EPA Method #353.3	0.01 to 1.0 mg NO3-N/L
Total Phosphorus (mg/l as P)	Grab Sample	EPA Method #365.3	0.01 to 1.2 mg P/L
Copper (mg/l)	Grab Sample	EPA Method #220.1	Low detection limit is 0.1 ppm
Iron (mg/l)	Grab Sample	EPA Method #236.1	Low detection limit is 0.1 ppm
Manganese (mg/l)	Grab Sample	EPA Method #243.1	Low detection limit is 0.1 ppm
Zinc (mg/l)	Grab Sample	EPA Method #289.1	Low detection limit is 0.1 ppm
Conductivity (mS/cm)	*In situ* Measurement		1 to 100 mS/cm
Dissolved Oxygen (%)	*In situ* Measurement		0% to 100%
Dissolved Oxygen (mg/l)	*In situ* Measurement		0.2 to 18.8 mg/L
PH	*In situ* Measurement		0 to 14 units
Turbidity (NTU)	*In situ* Measurement		5 to 1000 NTU
Water Temperature (°C)	*In situ* Measurement		−5°C to 50°C
Alkalinity (mg/l as CaCO3)	Grab Sample	EPA Method #310.1	All concentration ranges of alkalinity
Hardness (mg/l as CaCO3)	Grab Sample	EPA Method #130.2	All concentration ranges of hardness

Rothrock et al. (1998). A composite normalized metric (CNM) is calculated by dividing each separate metric score by the largest score so metric scores would vary between zero and one, and then summing all of the metric scores into a single score for comparison between sites. The reciprocal of the metrics that become smaller with increased ecological integrity are used to calculate the CNM, so a higher CNM score is indicative of higher biologic integrity. Various metrics are available for comparison. We have found that metrics suggested by Rothrock et al. (1998) (based upon a general set of metrics recommended by Plafkin et al., 1989) and metrics chosen by Stribling et al. (1998) will adequately assess the biological integrity of candidate sites (Table 4.6).

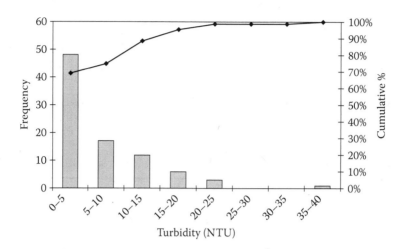

FIGURE 4.3 Frequency and cumulative percentage of turbidities of water at GIS-selected reference sites.

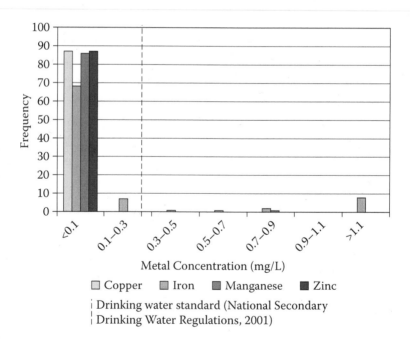

FIGURE 4.4 Frequency of metal concentrations in water samples from GIS-selected reference sites.

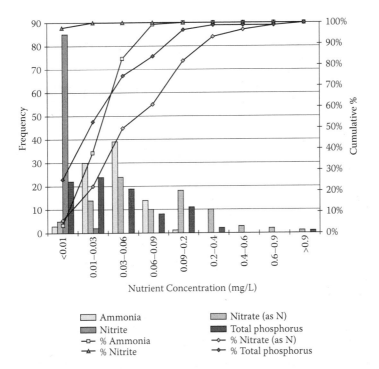

FIGURE 4.5 Frequency and cumulative percentage of nutrient concentrations in water samples from GIS-selected reference sites.

GEOGRAPHIC INFORMATION SYSTEM (GIS)-SELECTED REFERENCE SITES OR BEST PROFESSIONAL JUDGMENT (BPJ)?

The method described in the previous section is a relatively unbiased approach to choosing reference sites and does require an additional step of ground-truthing and analysis. Many regulatory agencies have chosen to rely upon their own staff or consultant scientists to provide best professional judgment to more rapidly enter the classification and prioritization stage of assessing stream health. Based upon our analysis of BPJ sites and GIS-selected sites in Georgia, we suggest that the unbiased GIS approach represents a more comprehensive description of the reference condition. We summarize our analysis of the comparison of BPJ and GIS sites in the following.

Three sets of previously identified candidate reference sites were examined: one set of 35 sites selected using best professional judgment by the EPA, Region 4, Science and Ecosystems Support Division; one set of 19 sites selected by best professional judgment for the Georgia Environmental Protection Division (11 overlapping with those provided by EPA); and a set of 70 sites, of which 35 were also identified by EPA and Georgia Environmental Protection Division (GAEPD), recommended by the Wildlife Resources Division of the Georgia Department of Natural Resources, based on fish IBI data.

The sites selected by BPJ underwent an abbreviated version of the same process used to select sites by GIS. Catchments were first delineated from the location of each site in the point theme created from the original lists. Catchments that had also

TABLE 4.6

Summary of Metrics Used in Characterizing Ecological Integrity

Source	Metric	Ecological Relevance
Rothrock–MBSS	Taxa Richness	The richness of the community indicates biodiversity of ecosystems and is used as a quantitative measure of stream water and habitat quality. Taxa richness generally decreases as a stream ecosystem degrades.
	EPT Richness	The richness of the intolerant insect orders Ephemeroptera (mayflies), Plecoptera (stoneflies), and Trichoptera (caddisflies) can indicate stream condition, since they tend to become scarcer with increasing disturbance.
Rothrock et al. (1998)	Total Abundance	Total number of organisms sampled can have variable response to stream impairment, but will generally decrease as a stream ecosystem degrades.
	EPT Abundance	The number of the these generally intolerant insects can indicate stream condition, since the organisms themselves also tend to become more scarce with increasing levels of disturbance.
	Percent Dominant Taxa	The proportion of the entire community composed of the most abundant taxa can be used to examine community balance. A community dominated by relatively few taxa is indicative of stress.
	Biotic Index	An index based on the tolerance of different organisms to pollution and stress. Tolerance values were based on values developed by Lenat (1993) or given in the RBP manual (Barbour et al. 1999).
	EPT:Chironomid	The ratio of EPT to Chironomids (midges). Decreasing ratios indicate stress since Chironomids tend to increase with increasing organic enrichment.
	Shredders:Total	Measure of distribution among functional feeding groups; shredders will decrease due to riparian zone impacts and can be indicators of toxins.
	Scrapers to Filterers	Shifts in functional feeding group between the scrapers who increase with increasing diatom abundance and filterers who increase with increasing filamentous algae indicate an over abundance of certain food sources.
MBSS (Stribling et al. 1998)	Number of Ephemeroptera Taxa	The richness of mayfly taxa indicates the ability of a stream to support these intolerant insects. Organic enrichment and excess fine sediment will often reduce the diversity of mayflies.
	Number of Diptera Taxa	Diptera as an order are relatively diverse and Dipterans are variable in their tolerance to stress. However, a high diversity of Diptera taxa generally suggests good water and habitat quality.

(Continued)

TABLE 4.6 (CONTINUED)
Summary of Metrics Used in Characterizing Ecological Integrity

Source	Metric	Ecological Relevance
MBSS (Stribling et al. 1998) (continued)	Percent Ephemeroptera	The dominance of the community by mayflies can indicate the relative success of these pollution intolerant individuals in sustaining reproduction. Stresses will reduce the abundance of mayflies relative to others.
	Number of Intolerant Taxa	Intolerant taxa are the first to be eliminated by perturbations, since they are often specialists with specialized habitat or water quality requirements. Taxa with tolerance ratings from 0 to 3 were considered intolerant.
	Percent Tolerant	As perturbation increases, tolerant individuals (tolerance values 7 to 10) tend to predominate in the sample. Intolerant individuals become less abundant as stress increases, leading to more individuals in tolerant taxa.
	Percent Tanytarsini of Chironomid	The tribe Tanytarsini is a relatively intolerant group of midges. The degree to which they represent the total number of midges indicates the general sensitivity of the midge assemblage. A high percentage of Tanytarsini among the midges may indicate lower levels of anthropogenic stress.
	Percent Collectors	Abundance of detritivores typically decreases with increased disturbance. This ecological response may be a food web effect, where organic material becomes scarce or unsuitable with increased perturbation.

been chosen by the GIS were eliminated and remaining catchments were evaluated to ensure they met the size standards for reference sites (within an order of magnitude of all of the sites for that subecoregion) and were located at least 80% within a single subecoregion. Land use measures were then made of each remaining catchment. These measurements were then compared to measurements of the catchments that had been selected as possible candidates by GIS.

The results of the analysis of the possible candidate reference sites chosen by BPJ are presented in Table 4.7. These sites only encompassed 15 of the 23 different subecoregions that we analyzed across the state. Of these, only sites in 5 subecoregions were sampled, with sites in the remaining 10 showing signs of impairment. For these 5 subecoregions, 14 GIS sites were then compared with 9 BPJ sites. Both nutrient and metal concentrations in these five subecoregions were very low, with all of the measurements except one below 1 mg/L. All sites, both GIS and BPJ, met the two standards set by the EPA for the nutrients nitrate and nitrite. In general, nutrient concentrations for GIS sites were equal to or less than the BPJ sites in all but 5 out of the 14 sites (see Figure 4.6 for results from subecoregions 45a–c, results from subecoregions 65c,g are not presented).

Whereas all of the GIS sites received suboptimal scores (ranging from 113 to 165) for the visual habitat assessment, four of the nine BPJ sites received optimal scores

TABLE 4.7
Results of Analysis of Best Professional Judgment Sites

Source	Number of Sites	Number of Subecoregions	Number Also Chosen by GIS	Number Rejected for Land Use	Number Rejected for Size	Number Rejected for Subecoregion Overlap	Number Visited	Number Sampled	Number Impaired
EPA	35	11	8	1	4	3	15	7	8
Georgia EPD	8	5	3	0	0	0	2	1	1
Wildlife Resources	35	5	12	11	1	0	9	1	8
Georgia Department of Natural Resources									
Total	78	15*	23	12	5	3	26	9	17

* The total number of subecoregions where best professional judgment sites were selected is not additive, but includes overlap where various agencies selected different sites in the same ecoregion.

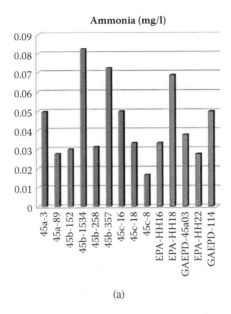

(a)

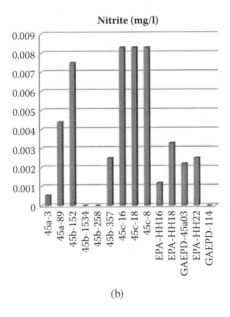

(b)

FIGURE 4.6 Comparisons of water quality at GIS-selected sites and BPJ sites in Ecoregion 45–Piedmont of Georgia. Sites EPA-HH16, EPA-HH18, GAEPD-45a03, EPA-HH22, and GAEPD-114 were selected using best professional judgment. The remaining nine sites were selected using GIS.

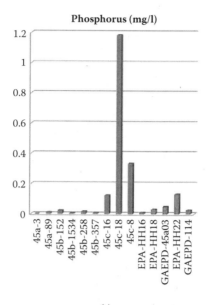

(c)

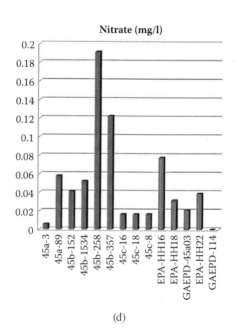

(d)

FIGURE 4.6   (*Continued*)

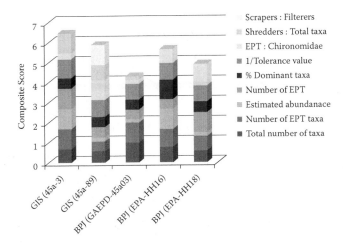

FIGURE 4.7 Comparison of composite macroinvertebrate metrics using Rothrock et al. (1998) criteria.

(166 to 200). BPJ sites were generally given higher scores than GIS sites, with means of 165 and 150, respectively. However, when viewed one subecoregion at a time, the GIS sites were not very different from the BJP sites, with the means being within 10 points of each other in three of the five cases.

The comparison of benthic macroinvertebrate communities using the general set of metrics used by Rothrock et al. (1998) is shown in Figure 4.7. Using this set of metrics, the GIS sites were either approximately equal to, or better than, the BPJ sites. However, the comparison is more equal when using the specific set of metrics used as part of the Maryland Biological Stream Survey (MBSS) (Stribling et al. 1998), shown in Figure 4.8. While three of the sites do not change significantly between the

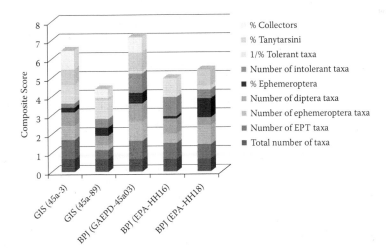

FIGURE 4.8 Comparison of composite macroinvertebrate metrics using MBSS and Stribling et al. (1998) criteria.

TABLE 4.8

Comparison of Candidate Sites against Maryland Reference Site Criteria

	Number of Sites Not Meeting Criteria		
Criteria	All Sites	GIS	BPJ
pH > 6 (if blackwater stream, ph < 6 and DOC < 8 mg/l)	5	5	0
Acid Neutralization Capability (ANC) > 50 meq/l	13	10	3
Dissolved O_2 > 4 ppm	4	3	1
Nitrate-N < 4.2 mg/l	0	0	0
Urban Land Use < 20% of Catchment Area	0	0	0
Forested Land Use > 25% of Catchment Area	0	0	0
In-Stream Habitat Optimal or Suboptimal	0	0	0
Riparian Buffer Width > 15 m	14	13	1
No Channelization	2	2	0
No Point Source Discharges	0	0	0
All Criteria	**32**	**27**	**5**
All Criteria, Except ANC	22	20	2
% Success, All Criteria	**63**	**65**	**44**
% Success, All Except ANC	75	74	78

Source: Stribling et al. (1998).

two metric sets, one of the GIS sites score is lowered while one of the BPJ sites score is increased. One of the causes of this change is the different ways of accounting for the presence/absence of tolerant organisms. Whereas the inverse of the tolerance value used by Rothrock et al. (1998) accounts for every organism in the subsample, the MBSS looked at only the most tolerant and intolerant organisms. The MBSS also did not use organism abundance as a measure.

The comparison of the candidate sites sampled in this study to the standards developed for reference sites as part of the MBSS is shown in Table 4.8. Sixty-five percent of the sites identified by GIS met all of Maryland's criteria for being a reference site. Since acid neutralization capability (ANC) can be a function of geology as much as a function of anthropogenic impact (Allen 1997), this criterion might not be applicable to Georgia. Approximately three-quarters of both the GIS and the BPJ sites met the remaining criteria.

CONCLUSIONS

If land use patterns analyzed by GIS are to be considered effective at identifying reference sites they must be able to identify enough of the least impaired sites available to provide reasonable statistical power and coverage of natural conditions in a cost-effective manner. GIS should also perform as well as, or better than, current methods of identifying reference sites based on local knowledge of experts. Judgment of the validity of the least impaired sites requires either data on all possible streams or a probabilistic sampling design to ascertain the current range of stream conditions available. While a final judgment on the method's ability to identify reference sites will have to await future studies of this type, currently available data indicate that the majority of streams identified by the unbiased GIS technique are not impaired.

Inevitably, some sites identified as having the least anthropogenic land use (the primary possible candidate reference sites) will be obviously impaired. The misidentification of sites probably stems from several sources. The fact that land use is constantly changing is one of the primary causes of misidentification. There is a lag time between publication of GIS data and many sites will suffer altered land use in the interim. Another cause of misidentification might be that moderate agricultural land use does not predict stream impairment in catchments with less than 50% agricultural land use. This low correlation of stream condition and agricultural land use is probably due to differences in land use that are below the resolution of the data available. This would include differences in agricultural practices such as fencing pastures from streams, contour plowing, and fallowing fields.

Water quality criteria for streams are generally based upon data from reference sites (EPA 2000), so judging whether the sites selected, using GIS, are "reference sites" using these criteria is a bit backward. However, EPA guidance does suggest using published threshold values for nutrient loading if no other resources exist. It does add, however, that most of these values are based on cobble bottom streams in northern temperate climates and these values should be used with caution in sandy bottomed or Southeastern streams. As shown in Table 4.9, published values also cover a wide range of thresholds for nutrient impairment. Using even the most restrictive standard for total phosphorus (less than 0.02 mg/L), a significant percentage (between 40% and 60%) of GIS sites might be classified as unimpaired.

It is also useful to compare the streams selected by GIS as candidates against those chosen by BPJ to determine if the GIS method is selecting sites as well. The GIS method has an immediate advantage over the BPJ method in its ability to evenly cover all portions of the state, whereas the BPJ method may fail to recommend sites in some subecoregions. Both methods will select the relatively unimpaired streams, if they are available. Our water chemistry data from Georgia subecoregions with sites selected by both methods supports this conclusion, with

TABLE 4.9
Published Literature Threshold Values for Nutrient Impairment

Source	Nitrogen (mg/L)	Type of Measurement	Phosphorus (mg/L)	Type of Measurement
Dodds et al. 1998, cited in EPA 2000	1.500	Total N	0.075	Total P
Dodds et al. 1997, cited in EPA 2000	0.27–0.65	Total N	0.038–0.090	Total P
Tennessee Reference Site Maximum Values, EPA 2000	2	NO_{2+3}	0.26	Total P
Van Nieuwenhuyse and Jones 1996, cited in EPA 2000	0.3	Total N	0.042	Total P
Lenat and Crawford 1994, mean values for forested Piedmont streams in North Carolina	0.08	NO_{2+3}	0.09	Total P
Lenat and Crawford 1994, mean values for agricultural Piedmont streams in North Carolina	0.59	NO_{2+3}	0.27	Total P
Chetelat et al. 1999, cited in EPA 2000			0.02	Total P
Dodds and Welch 2000	0.47	Total N	0.06	Total P

all the measurements of potential nutrient contaminants being low. In some subecoregions the GIS sites had lower concentrations, and in others the BPJ sites had lower concentrations; the number of each was approximately equal.

The conclusion that the GIS sites are less impaired than or equally impaired as the BPJ sites is also supported by the biological data. Whether GIS sites are better than BPJ sites depends upon which set of metrics are utilized. It will be necessary to develop appropriate metrics that reflect the unique physical and chemical conditions within each ecoregion or subecoregion. For example, sites in Georgia did not score well when using the MBSS metrics, which rely upon the number and diversity of Ephemeroptera, probably because of naturally low Ephemeropteran richness in the southeastern Unites States. The metric of number of intolerant taxa used in the normalized scores is also somewhat deceiving. When few intolerant taxa are present, the addition of a single individual of another intolerant taxon can have significant impact on the composite metric score. The choice of collectors as a measure of functional feeding group composition is also somewhat problematic. Stribling et al. (1998) noted that while the percentage of collectors did work well for identifying stress in noncoastal plain sites in Maryland, the coastal plain sites had a reciprocal relationship with the percentage of collectors increasing with stress.

In addition to the ability to identify the least impacted streams available, the unbiased GIS method chosen to select the candidate reference sites method should also minimize time and cost. This analysis required a single desktop computer, a single GIS analyst, and approximately three months of CPU and "man-time," including the time for method development. Although this method requires a greater amount of effort and expertise, it does deliver a product of predictable quality, unlike relying upon local expertise, which may vary greatly from place to place. In the future, this method may become even more reliable. We recommend the following changes/additions to the method:

1. Analysis should be done at multiple scales. We recommend a two- or three-category approach, analyzing catchments in the 5 to 50 km² and the 50 to 500 km² ranges. This would both increase the number of sites located and increase the amount of natural variation covered by the sites to include differences in size.
2. The most recent land use data should be used, if available. For rapidly changing landscapes, up-to-date land use data will improve the proportion of sites identified as candidates ultimately being true reference sites.
3. A better way of estimating which land is in active silvicultural use should be developed and used for areas, like Georgia, that have significant portions of land in silviculture. Ideally, a data set describing silvicultural land use at a relatively high resolution may someday be available (currently data on silviculture are only available for county size units and larger). If not, a better estimate of how much land is in silviculture may be made by combining barren land and land that is exclusively evergreen forest, since both of these land types are predominate in silvicultural areas and relatively rare in natural areas.
4. Use a method of scoring sites based on their distributions such as a z-score. The z-score is calculated by normalizing all data to the mean, so scores below the mean are negative and those above the mean are positive. These scores can then be summed and weighted as appropriate, and used to rank sites.

The GIS method has several advantages over identifying sites using BPJ. It is an objective method of identifying potential reference sites instead of relying on the subjective judgment of experts or by trying to extrapolate from limited data. The ability to rank all streams in a subecoregion provides a greater number of potential reference sites with a wider geographic distribution. This allows more sites to be sampled over a greater range of conditions to better define the reference condition. This method also assesses all streams, not just those that have relatively easy access as is done with sites selected using BPJ. Access is gained to sites that are assessed by professionals usually by public road. However, the presence of public roads in the catchment is also a form of disturbance, so the professionals choosing sites do not generally assess the least disturbed sites, those with the most difficult access.

The protection and restoration of aquatic ecosystems depends on having a clear defensible standard of what is attainable both chemically and biologically. The best method of defining this standard is by using a reference condition based upon a set of least impaired reference sites that represent the range of natural conditions present. The quality of the bioassessments based upon this reference condition ultimately is a function of the quality of the sites used to define this reference condition. States or other regulatory entities that must establish reference sites can use this method to rapidly select a set of potential reference sites. States with a currently designated set of reference sites can use this unbiased GIS method to add additional sites to better define the reference condition.

REFERENCES

Abbruzzese, B., and S.G. Leibowitz. 1997. A synoptic approach for assessing cumulative impacts to wetlands. *Environmental Management* 21: 457–475.

Allen, J.D. 1997. *Stream Ecology: Structure and Function of Running Waters.* Philadelphia: Chapman & Hall.

Bailey, R.C., M.G. Kennedy, M.Z. Dervish, and R.M. Taylor. 1998. Biological assessment of freshwater ecosystems using a reference condition approach: Comparing predicted and actual benthic invertebrate communities in Yukon streams. *Freshwater Biology* 39: 765–774.

Barbour, M.T., J. Gerritsen, B.D. Snyder, and J.B. Stribling. 1999. *Rapid Bioassessment Protocols for Use in Streams and Wadeable Rivers: Periphyton, Benthic Macroinvertebrates and Fish*, 2nd ed. EPA 841-B-99-002. Washington, DC: U.S. Environmental Protection Agency, Office of Water.

Bolstad, P.V. and W.T. Swank. 1997. Cumulative impacts of landuse on water quality in a Southern Appalachian watershed. *Journal of the American Water Resources Association* 33(3): 519–533.

Bryce, S.A., and S.E. Clarke. 1996. Landscape-level ecological regions: Linking state-level ecoregion frameworks with stream habitat classifications. *Environmental Management* 20(3): 297–311.

Chetelat, J., F.R. Pick, and A. Morin. 1999. Periphyton biomass and community composition in rivers of different nutrient status. *Canadian Journal of Fisheries and Aquatic Sciences* 56(4): 560–569.

Clean Water Act of 1977, § 101(a), 33 U.S.C., § 1251(a) (1999).

Clean Water Act of 1977, § 303(c)(2)(B), 33 U.S.C., § 1313(c)(2)(B) (1999).

Clean Water Act of 1977, § 305(b)(1)(B), 33 U.S.C., § 1315(b)(1)(B) (1999).

Clean Water Act of 1977, § 319, 33 U.S.C., § 1329 (1999).

Dodds, W.K., and E.B. Welch. 2000. Establishing nutrient criteria in streams. *Journal of the North American Benthological Society* 19(1): 186–196.

Dodds, W.K., J.R. Jones, and E.B. Welch. 1998. Suggested classification of stream trophic state: Distributions of temperate stream types by chlorophyll, total nitrogen, and phosphorus. *Water Research* 32(5): 1455–1462.

Dodds, W.K., V.H. Smith, and B. Zander. 1997. Developing nutrient targets to control benthic chlorophyll levels in streams: A case study of the Clark Fork River. *Water Research* 31(7): 1738–1750.

Ebert, D.W., and T.G. Wade. 1999. *Analytical Tools Interface for Landscape Assessments (ATtILA) User Guide (ArcView GIS extension)*, version 2.0. Las Vegas, NV: U.S. Environmental Protection Agency, Office of Research and Development, National Exposure Research Laboratory, Environmental Sciences Division, Landscape Ecology Branch.

Fore, L.S., J.R. Karr, and R.W. Wisseman. 1996. Assessing invertebrate response to human activities: Evaluating alternative approaches. *Journal of the North American Benthological Society* 15(2): 212–231.

Gallant, A.L., T.R. Whittier, D.P. Larsen, J.M. Omernik, and R.M. Hughes. 1989. *Regionalization as a Tool for Managing Environmental Resources.* EPA-600-3-89-060. Corvallis, OR: U.S. Environmental Protection Agency, Environmental Research Laboratory.

Gibson, G.R., M.T. Barbour, J.B. Stribling, J. Gerritsen, and J.R. Karr. 1996. *Biological Criteria: Technical Guidance for Streams and Small Rivers* (rev. ed.). EPA 822-B-96-001. Washington, DC: U.S. Environmental Protection Agency, Office of Science and Technology.

Hooge, Phillip N. 1998. Spatial Tools [ArcView extension on the Internet]. Version 3.3. Anchorage, AK: U.S. Geologic Survey, Biological Resource Division. [Accessed: January–February 2000]. Available from: http://www.absc.usgs.gov/glba/gistools/spatialtools_doc.htm.

Hornig, C.E., C.W. Bayer, S.R. Twidwell, J.R. Davis, R.J. Kleinsasser, G.W. Linam, and K.B. Mayes. 1995. Development of regionally based biological criteria in Texas. In *Biological Assessment and Criteria: Tools for Water Resource Planning and Decision Making*, eds. W.P. Davis and T.P. Simon, 145–147. Boca Raton, FL: CRC Press.

Hughes, R.M. 1995. Defining acceptable biological status by comparing with reference conditions. In *Biological Assessment and Criteria: Tools for Water Resource Planning and Decision Making*, eds. W.P. Davis and T.P. Simon, 31–47. Boca Raton, FL: CRC Press.

Hughes, R.M., D.P. Larsen, and J.M. Omernik. 1986. Regional reference sites: A method for assessing stream potentials. *Environmental Management* 10(5): 629–635.

Johnson, L.B., C. Richards, G.E. Host, and J.W. Arthur. 1997. Landscape influences on water chemistry in Midwestern stream ecosystems. *Freshwater Biology* 37: 193–208.

Johnson R.K., T. Wiederholm, and D.M. Rosenberg. 1993. Freshwater biomonitoring using individual organisms, populations, and species assemblages of benthic macroinvertebrates. In *Freshwater Biomonitoring and Benthic Macroinvertebrates,* eds. D.M. Rosenberg and V.H. Resh, 40–158. New York: Chapman & Hall.

Kennen, J.G. 1999. Relation of macroinvertebrate community impairment to catchment characteristics in New Jersey streams. *Journal of the American Water Resources Association* 35: 939–955.

Lammert, M., and J.D. Allen. 1999. Assessing biotic integrity of streams: Effects of scale in measuring the influence on land use/cover and habitat structure on fish and macroinvertebrates. *Environmental Management* 23: 257–270.

Lenat, D.R. 1993. A biotic index for the southeastern United States: Derivation and list of tolerance values, with criteria for assigning water-quality ratings. *Journal of the North American Benthological Society* 12(3): 279–290.

Lenat, D.R., and J.K. Crawford. 1994. Effects of land use on water quality and aquatic biota of three North Carolina Piedmont streams. *Hydrobiologia* 294: 185–199.

Omernik, J.M. 1987. Map supplement: Ecoregions of the conterminous United States. *Annals of the Association of American Geographers* 77: 118–125.

Omernik, J.M. 1995. Ecoregions: A spatial framework for environmental management. In *Biological Assessment and Criteria: Tools for Water Resource Planning and Decision Making*, eds. W.P. Davis and T.P. Simon, 49–62. Boca Raton, FL: CRC Press.

Plafkin, J.L., M.T. Barbour, K.D. Porter, S.K. Gross, and R.M. Hughes. 1989. *Rapid Bioassessment Protocols for Use in Streams and Rivers: Benthic Macroinvertebrates and Fish*. EPA 440-4-89-001. Washington, DC: U.S. Environmental Protection Agency, Assessment and Watershed Protection Division.

Resh, V.H., R.H. Norris, and M.T. Barbour. 1995. Design and implementation of rapid assessment approaches for water resource monitoring using benthic macroinvertebrates. *Australian Journal of Ecology* 20: 108–121.

Reynoldson, T.B., R.H. Norris, V.H. Resh, K.E. Day, and D.M. Rosenberg. 1997. The reference condition: A comparison of multimetric and multivariate approaches to assess water-quality impairment using benthic macroinvertebrates. *Journal of the North American Benthological Society* 16(4): 833–852.

Richards, C., R.J. Haro, L.B. Johnson, and G.E. Host. 1997. Catchment and reach-scale properties as indicators of macroinvertebrate species traits. *Freshwater Biology* 37: 219–230.

Richards, C., and G.E. Host. 1994. Examining land use influences on stream habitats and macroinvertebrates: A GIS approach. *Journal of the American Water Resources Association* 30(4): 729–738.

Richards, C., L.B. Johnson, and G.E. Host. 1996. Landscape-scale influences on stream habitats and biota. *Canadian Journal of Fisheries and Aquatic Sciences* 53(Suppl. 1): 295–311.

Roth, N.E., J.D. Allen, and D.L. Erickson. 1996. Landscape influences on stream biotic integrity assessed at multiple spatial scales. *Landscape Ecology* 11(3): 141–156.

Roth, N.E., M.T. Southerland, G. Mercurio, J.C. Chaillou, P.F. Kazyak, S.S. Stranko, A.P. Prochaska, D.G. Heimbuch, and J.C. Seibel. 1999. *State of the streams: 1995–1997 Maryland biological stream survey results.* Report CBWP-MANTA-EA-99-6. Annapolis: Maryland Department of Natural Resources, Monitoring and Non-Tidal Assessment Division.

Rothrock, J.A., P.K. Barten, and G.L. Ingman. 1998. Land use and aquatic biointegrity in the Blackfoot River watershed, Montana. *Journal of the American Water Resources Association* 34(3): 565–581.

Schnackenberg, E.S., and L.H. MacDonald. 1998. Detecting cumulative effects on headwater streams in the Routt National Forest, Colorado. *Journal of the American Water Resources Association* 34(5): 1163–1177.

Science Advisory Board (SAB). 1993. *Review of Draft Technical Guidance for Biological Criteria for Streams and Small Rivers.* EPA-SAB-EPEC-94-003. Washington, DC: U.S. Environmental Protection Agency, Science Advisory Board.

Stanford, J.A. 2006. Landscapes and riverscapes. In *Methods in Stream Ecology*, 2nd ed., eds. F.R. Hauer and G.A. Lamberti, 3–21. San Diego, CA: Academic Press.

Strahler, A.N. 1952. Hypsometric (area-latitude) analysis of erosional topography. *Bulletin of the Geological Society of America* 63: 1117–1142.

Stribling, J.B., B.K. Jessup, J.S. White, D. Boward, and M. Hurd. 1998. *Development of a Benthic Index of Biotic Integrity for Maryland Streams.* CBWP-EA-98-3. Annapolis: Maryland Department of Natural Resources, Monitoring and Non-Tidal Assessment Division.

U.S. Environmental Protection Agency, Office of Water. 1998a. *EPA/OW Industrial Facilities Discharge Database for CONUS* [locational point file in ArcView shapefile vector format on the Internet]. Washington, DC: U.S. Environmental Protection Agency. http://www.epa.gov/ost/basins/ (accessed April 2000).

U.S. Environmental Protection Agency, Office of Water. 1998b. *EPA/OW Permit Compliance System for CONUS* [locational point file in ArcView shapefile vector format on the Internet]. Washington, DC: U.S. Environmental Protection Agency. http://www.epa.gov/ost/basins/ (accessed April 2000).

U.S. Environmental Protection Agency, Office of Water. 2000. *Nutrient Criteria Technical Guidance Manual Rivers and Streams.* EPA 822-B-00-002. Washington, DC: U.S. Environmental Protection Agency.

Van Nieuwenhuyse, E.E., and J.R. Jones. 1996. Phosphorus–chlorophyll relationship in temperate streams and its variation with stream catchment area. *Canadian Journal of Fisheries and Aquatic Sciences* 53(1): 99–105.

Vogelmann, J.E., S.M. Howard, L. Yang, C.R. Larson, B.K. Wylie, and N. Van Driel. 2001. Completion of the 1990s national land cover data set for the conterminous United States from Landsat Thematic Mapper data and ancillary data sources. *Photogrammetric Engineering & Remote Sensing* 67(6): 650–662.

Wang, L., J. Lyons, P. Kanehl, and R. Gatti. 1997. Influences of watershed land use on habitat quality and biotic integrity in Wisconsin streams. *Fisheries* 22(6): 6–12.

Warry, N.D., and M. Hanau. 1993. The use of terrestrial ecoregions as a regional-scale screen for selecting representative reference sites for water quality monitoring. *Environmental Management* 17(2): 267–276.

5 Development of Ecoregional and Subecoregional Reference Conditions

Duncan L. Hughes, John R. Olson,
Michele P. Brossett, and James A. Gore

CONTENTS

Reference conditions are a set of physical, chemical, and biological conditions that describe the characteristics of the most minimally impaired streams in a given ecoregion or subecoregion. Although derived from a set of measurements acquired from typical reference stream sites in a relatively homogeneous region, the reference condition is based upon a composite of those characteristics and is not, therefore, site specific. An established reference condition provides a basis for making comparisons and making determinations on impairment or degree of impairment for a newly sampled stream in that same ecoregional unit (Gibson et al. 1996).

Ideally, all reference conditions would be based upon evaluation of as many unimpaired streams in any given ecoregion or subecoregion. However, it is likely that no pristine, truly unimpaired streams exist in many regions (Minshall 1988). Thus, the reference condition in each ecoregion and subecoregion is based upon evaluation of the least impaired or minimally impaired streams in the region, as noted in the choice of candidate reference sites. The presentation of reference conditions, then, is based upon the assumption that the least impaired streams are evaluated during the same index periods and that future samples from streams for comparison to the reference condition will also be taken during that same index period. For example, in the state of Georgia, we chose an index period of September through February. This period was chosen because understory development is minimal, making access to the stream easier. In addition, exposure to disease vectors (ticks, mosquitoes, and so forth) was also reduced and evaluation of physical habitat structure was optimized. In addition, this is a relatively stable time interval, which allowed several months to sample a large number of streams containing larger macroinvertebrate individuals (later instars) yet without the impact of emergence of adult aquatic insects, where community composition might be erroneously reported.

The reference condition, being based on minimally impaired streams, should be continuously updated and reevaluated through time. Candidate reference sites should be revisited and sampled and a geographic information system (GIS) evaluation of each ecoregion performed, with updated data layers. By using the same rules from which we chose our reference conditions, management and regulatory agencies should be able to create timely and accurate comparisons to other streams in the same ecoregion or subecoregion.

PRELIMINARY REFERENCE CONDITION CRITERIA

Initial evaluations of candidate reference streams are based upon abiotic factors (land use, physical habitat, and water chemistry). Preliminary thresholds for chemical parameters were obtained from published values within federal and state surface water quality guidelines. Land use data, physical habitat assessment scores, and *in situ* and laboratory chemistry values are plotted for each ecoregion and subecoregion. Box-and-whisker plots are evaluated to establish percentile thresholds for similar streams. A comparison of preliminary target levels with actual data further refines the quantitative threshold for each abiotic factor plotted. A candidate reference stream must meet the preliminary reference condition threshold criteria established for the ecoregion or subecoregion (hereafter, "site groups") to be considered for use in defining the biological condition of the characteristic reference stream. These criteria have been presented in Chapter 4.

BIOLOGICAL METRICS AND THE MULTIMETRIC APPROACH

Streams meeting the preliminary criteria for reference condition are evaluated using a multimetric approach to assess the integrity of the biological community. Metrics are biological attributes of the benthic community that indicate ambient water quality conditions. Metrics attempt to quantify aspects of the structure and function of the benthic community and may be divided into five major groups: (1) taxonomic richness, (2) community composition, (3) comparative assessments of tolerant and intolerant species in the assemblage, (4) proportions of members of various functional feeding groups, and (5) proportions assigned to various feeding and foraging habits.

Taxonomic Richness

These metrics describe the numbers of distinct taxa within taxonomic groups (for example, total taxa, numbers of EPT [Ephemeroptera, Plecoptera, and Trichoptera] taxa, or numbers of Diptera taxa). High taxonomic richness usually correlates with better water quality and stream health.

Community Composition

Composition metrics are usually expressed as percentages, and indicate the proportion of individuals in a sample belonging to a specific taxonomic group. Some composition measures may also serve as tolerance/intolerance metrics (for example, percentage of *Chironomus* spp. and *Cricotopus* spp./total Chironomidae) where certain families or genera have an established higher tolerance to pollution than the other members of the order or family, respectively.

Tolerance/Intolerance

Tolerance metrics represent the general level of tolerance to pollution (most often, to organic loading) by biota within a sample. Some are weighted scores based on tolerance classes (e.g., Beck's Index; Beck 1965), and some are based on the average tolerance values of individuals within the sample (e.g., the North Carolina Biotic Index; Lenat 1993).

Functional Feeding Group

These metrics indicate dominant feeding mechanisms of the biological assemblage. Some specialized feeders are more sensitive to disturbance and pollution than more generalized feeders (Rosenberg and Resh 1996).

Habit

Habit metric categories include taxonomic richness and composition measures describing movement and positioning mechanisms of benthic organisms (for example, "swimmer" taxa or percent "sprawlers") (Merritt and Cummins 1996).

Combining individual metrics into a multimetric index allows integration of different indicators into a single ecologically based index. Approximately 70 different biological metrics can be evaluated. With this large body of potential metrics for each ecoregion or subecoregion, a computer-based spreadsheet system should be used for these calculations. For example, in the state of Georgia, we used the Ecological Data Application System (EDAS), version 3.3.2k, to calculate metrics for each reference site. Only metrics with an established response to stress (e.g., from Rapid Bioassessment Protocol [RBP] or other literature) were used in final index development.

TABLE 5.1

Predicted Responses of Benthic Macroinvertebrate Metrics to Stress

Metric Category	Metric	Stress Response
Richness	Total Taxa	Decrease
	Ephemeroptera, Plecoptera, and Trichoptera (EPT) Taxa	Decrease
	Ephemeroptera Taxa	Decrease
	Plecoptera Taxa	Decrease
	Trichoptera Taxa	Decrease
	Coleoptera Taxa	Decrease
	Diptera Taxa	Decrease
	Chironomidae Taxa	Decrease
	Tanytarsini Taxa	Decrease
	Evenness	Decrease
	Margalef's Index	Decrease
	Shannon's Index (base e)	Decrease
	Simpson's Diversity Index	Increase
Composition	% EPT	Decrease
	% Ephemeroptera	Decrease
	% Amphipoda	Decrease
	% Chironomidae	Increase
	% Coleoptera	Decrease
	% Diptera	Increase
	% Gastropoda	Decrease
	% Isopoda	Increase
	% Noninsect	Increase
	% Odonata	Increase
	% Plecoptera	Decrease
	% Tanytarsini	Decrease
	% Oligochaeta	Increase
	% Trichoptera	Decrease
	% Chironominae/Total Chironomidae (TC)	Variable
	% Orthocladiinae/TC	Decrease
	% Tanypodinae/TC	Increase
	% Hydropsychidae/Total Trichoptera	Increase
	% Hydropsychidae/Total EPT	Increase
	% Tanytarsini/TC	Decrease
	% *Cricotopus* sp. and *Chironomus* sp./TC	Increase
Tolerance/Intolerance	Tolerant Taxa	Increase
	% Tolerant Individuals	Increase
	Intolerant Taxa	Decrease
	% Intolerant Individuals	Decrease
	% Dominant Individuals	Increase
	Dominant Individuals	Increase

TABLE 5.1 (CONTINUED)
Predicted Responses of Benthic Macroinvertebrate Metrics to Stress

Metric Category	Metric	Stress Response
	Beck's Index	Decrease
	Hilsenhoff's Biotic Index (HBI)	Increase
	North Carolina Biotic Index (NCBI)	Increase
Functional Feeding Group	% Scraper	Decrease
	Scraper Taxa	Decrease
	% Collector	Decrease
	Collector Taxa	Decrease
	% Predator	Decrease
	Predator Taxa	Decrease
	% Shredder	Decrease
	Shredder Taxa	Decrease
	% Filterer	Increase
	Filterer Taxa	Decrease
Habit	Clinger Taxa	Decrease
	% Clinger	Decrease
	Burrower Taxa	Decrease
	Climber Taxa	Decrease
	Sprawler Taxa	Decrease
	Swimmer Taxa	Decrease

Other factors to be considered for candidate metric evaluation include:

- Importance within the ecoregion or subecoregion under examination
- Low incremental cost
- Responsiveness to stressors on a regional scale
- Method of measurement is feasible on a regional scale

A list of candidate metrics, metric category, and response to stress is included in Table 5.1.

BIOTIC INDEX DEVELOPMENT

A valid metric should distinguish reference from impaired streams. Thus, final reference criteria must be refined and calibrated against data collected from known impaired streams. Impaired sites can be chosen using the same method described for choosing candidate reference sites in Chapter 4. Those streams that do not qualify as candidate reference sites are, by default, impaired to one level or another. From this second grouping of streams, samples for comparison to the reference condition can be obtained. Figure 5.1 and Figure 5.2 depict the comparison between reference and impaired sites in the state of Georgia.

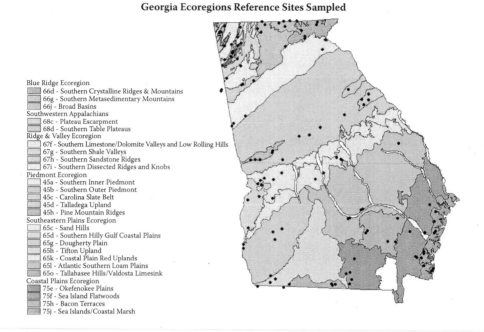

FIGURE 5.1   Georgia ecoregions reference stream sites.

FIGURE 5.2   Georgia ecoregions and impaired sites sampled.

Raw metric scores for reference and impaired streams in each site group are used to calculate discrimination efficiency of each metric. Discrimination efficiency (DE) is calculated with the formula

$$DE = 100 \times a/b,$$

where a is the number of impaired streams scoring below the 25th percentile of the reference distribution for metrics that decrease with stress (or the number of impaired streams scoring above the 75th percentile of the reference distribution for metrics that increase with stress), and b is the total number of impaired samples.

Box-and-whisker plots of site group distributions (of reference and impaired stream classes) for metrics with the highest DE in each metric category are also evaluated. The best performing metrics minimizes variation within stream classes and maximizes variation among stream classes.

Pearson product-moment correlation analyses of metrics will reveal those redundant metrics (Glantz 1992). All metrics that do not differentiate reference from impaired streams are removed from consideration for inclusion in the candidate composite index for that particular ecoregion or subecoregion. Redundant metrics are removed from consideration for inclusion in the same candidate index. To increase the sensitivity of the final metric index and assure that all response variables are considered, every effort should be made to assure that at least one metric from each of the five metric categories is included in the final composite metrics. The individual metrics from each metric category that best differentiate reference streams from impaired streams, and meet the aforementioned criteria, are considered for inclusion in candidate indices.

After elimination of redundant and low discrimination metrics, the remaining candidate metrics raw data values are standardized for comparison to one another. The method of standardization varies, depending on whether the metric score increases or decreases in response to stress.

For metrics that decreased with stress (e.g., EPT taxa):

$$\text{Standardized Score} = 100 \times c/d,$$

where c is the metric value for a test stream within a site group, and d is the 95th percentile value of the reference stream distribution for the site group.

For metrics that increased with stress (e.g., Hilsenhoff's Biotic Index):

$$\text{Standardized Score} = 100 \times [(e - c)/(e - f)],$$

where c is the metric value for a test stream within a site group, e is the highest observed value among all streams within the site group, and f is the 5th percentile value of the reference stream distribution for the site group.

Standardized metrics from each metric category are then combined into candidate indices for each ecoregion or subecoregion. Composite indices are assembled in an additive manner:

$$\text{Composite Index Score} = (g + h + i + j + k + l + \cdots)/n,$$

where $g, h, i, j, k, l, \ldots$ is the standardized score of the best candidate metrics, and n is the total number of metrics included in the index. In most cases, final indices contain five to seven metrics, and are scored on a 0- to 100-point scale. The candidate index for each site group that provided the highest DE and best box-and-whisker plots is selected as the final index.

The multimetric approach assimilates biological data, with various ecological meanings, into a single index (for each stream group evaluated) to gauge the health of a stream. Final indices are comprised of five to seven metrics with at least one metric chosen from each of the metric categories (richness, composition, tolerance/intolerance, functional feeding group, and habit). No metric with a DE of less than 0.5 should be used in final index development. A summary of the composition of both ecoregional and subecoregional indices for the state of Georgia is provided in Table 5.2 through Table 5.6. An examination of these tables indicates that distinctive metrics characterize each ecoregion and subecoregion and that there is no common set of metrics that can describe the health of all wadeable streams.

In some areas (like the coastal plain), richness and habit metric categories do not differentiate reference streams from impaired streams (that is, with DE <0.5). These metric categories should be excluded from use in indices for these regions. A possible explanation is that individuals representing the biological attributes measured by the richness and habit measures were absent or minimal in both reference and impaired streams in these areas (e.g., EPT taxa when there are no EPTs present). In some regions (again, like coastal areas), many of the sample individuals are noninsect (e.g., amphipoda, isopoda, gastropoda, oligochaeta, and so forth) that are not well represented by traditional richness metrics. Also, the habits of many of these organisms are unknown; thus, undifferentiated between reference and impaired streams. Obviously, as these habits are elucidated by future research, the habit metrics may take on new meaning when recalculating future indices of stream health.

Where the RBP system has been applied on a regional or statewide basis, it seems that tolerance/intolerance metrics best distinguished reference from impaired streams (Stribling et al. 1998; Jessup and Gerritsen 2000; Mississippi Department of Environmental Quality [MDEQ] 2003; Gore et al. 2005). There are always some anomalous or unusual outcomes; in Georgia, four subecoregion indices used composition metrics that acted as surrogates of pollution tolerance (e.g., % Hydropsychidae/Total Trichoptera) as substitutes for traditional tolerance metrics (e.g., Hilsenhoff's Biotic Index [HBI], % Intolerant Individuals) that did not perform as well. It is noteworthy that the "universal metric," total taxa in the sample, has among the lowest overall average discrimination efficiencies (in Georgia, DE among subecoregions of 0.290) and is rarely included in any index. Summaries of the typical performance of individual metrics for Georgia ecoregions and subecoregions are contained in Table 5.7 and Table 5.8, respectively.

In addition, we include (Appendix E) containing some typical entries, from the Georgia Ecoregion Study (Gore et al. 2005), as this information is necessary for regulatory agencies, consultant environmental scientists, and researchers, as a basis of comparison, when evaluating the integrity of wadeable streams.

TABLE 5.2
Summary of Benthic Macroinvertebrate Indices for Major Ecoregions of Georgia

Metric Category	Index 45, Piedmont	Index 65, Southeastern Plains	Index 66, Blue Ridge	Index 67, Ridge and Valley	Index 68c&d, Southwestern Appalachians	Index 75, Southern Coastal Plain
Richness	EPT Taxa	EPT Taxa Margalef's Index	Plecoptera Taxa Simpson's Index	EPT Taxa Plecoptera Taxa	Plecoptera Taxa	
Composition	% Chironomidae	% Oligochaeta	% Trichoptera	% Plecoptera	% Hydropsychidae/Total Trichoptera	% Noninsect
	% Plecoptera % Odonata % EPT	% Tanypodinae/TC		% Isopoda	% Tanypodinae/TC	% Oligochaeta % Odonata % Tanypodinae/TC
Tolerance/ Intolerance	NCBI	% Intolerant Individuals	% Intolerant Individuals NCBI	HBI	NCBI	HBI
FFG		% Predator	Predator Taxa Burrower Taxa	Clinger Taxa	Scraper Taxa	
Habit		% Clinger			% Clinger	

TABLE 5.3

Summary of Macroinvertebrate Indices for Wadeable Streams in Subecoregions in the Piedmont (Ecoregion 45) in Georgia

Metric Category	Index 45a, Southern Inner Piedmont	Index 45b, Southern Outer Piedmont	Index 45c, Carolina Slate Belt	Index 45d, Talladega Upland	Index 45h, Pine Mountain Ridges
Richness	EPT Taxa	Coleoptera Taxa	Tanytarsini Taxa	Coleoptera Taxa	Plecoptera Taxa
Composition	% Chironomidae	% Oligochaeta	% Odonata	% Odonata	% Ephemeroptera
	% *Chironomus* and *Cricotopus*/TC	% Chironomidae	% Tanypodinae/TC	% Tanypodinae/TC	% Plecoptera
Tolerant/Intolerant	NCBI	% Intolerant Individuals	Dominant individuals	NCBI	% Intolerant Individuals
			% Intolerant Individuals	% Tolerant Individuals	
FFG	% Scraper	Scraper Taxa	% Shredder	Shredder Taxa	% Scraper
Habit	% Clinger	Swimmer Taxa	Swimmer Taxa		% Clinger

TABLE 5.4

Summary of Macroinvertebrate Indices for Wadeable Streams in Subecoregions in Ecoregion 65 in Georgia

Metric Category	Index 65d, Southern Hilly Gulf Coastal Plain	Index 65g, Dougherty Plain	Index 65h, Tifton Upland	Index 65k, Coastal Plain Red Uplands	Index 65l, Atlantic Southern Loam Plains
Richness	Plecoptera Taxa Trichoptera Taxa	EPT Taxa	Ephemeroptera Taxa		Diptera Taxa Trichoptera Taxa
Composition	% Oligochaeta	% Oligochaeta	% Isopoda % Tanytarsini	% Tanypodinae/TC % Gastropoda	% EPT
Tolerant/Intolerant	% Hydropsychidae/Trichoptera	HBI % Intolerant individuals	% Tolerant Individuals	% Hydropsychidae/Total Trichoptera	% Tolerant Individuals
FFG	% Predator % Filterer	Filterer Taxa	% Scraper	Scraper Taxa % Shredder % Collector	Shredder Taxa
Habit	Clinger Taxa	Clinger Taxa	Burrower Taxa		Clinger Taxa

TABLE 5.5

Summary of Macroinvertebrate Indices for Wadeable Streams in Subecoregions in Ecoregions 66 and 67 in Georgia

Metric Category	Index 66d, Southern Crystalline Ridges and Mountains	Index 66g, Southern Metasedimentary Mountains	Index 66j, Broad Basins	Index 67f&i, Southern Limestone/Dolomite Valleys and Low Rolling Hills; and Southern Dissected Ridges and Knobs	Index 67g, Southern Shale Valleys
Richness	Diptera Taxa	EPT Taxa	Simpson's Diversity Index Margalef's Index	Plecoptera Taxa Ephemeroptera Taxa	Plecoptera Taxa
Composition	% Plecoptera % Odonata	% Chironomidae % Tanypodinae/TC	% Tanytarsini	% EPT	% Orthocladiinae/TC
Tolerant/Intolerant	% Dominant Individuals	NCBI % Dominant Individuals	% Intolerant Individuals	NCBI Beck's Index	% Hydropsychidae/Total Trichoptera
FFG	% Shredder	Scraper Taxa	Predator Taxa	Scraper Taxa	Shredder Taxa Collector Taxa
Habit	Clinger Taxa	% Clinger	Sprawler Taxa	% Clinger	Sprawler Taxa

TABLE 5.6
Summary of Macroinvertebrate Indices for Wadeable Streams in Subecoregions in Ecoregions 68 and 75 in Georgia

Metric Category	Index 68c&d, Plateau Escarpment and Southern Table Plateaus	Index 75e, Okefenokee Plains	Index 75f, Sea Island Flatwoods	Index 75h, Bacon Terraces	Index 75j, Sea Islands/Coastal Marsh
Richness	Plecoptera Taxa		Chironomidae Taxa		
Composition	% Hydropsychidae/Total Trichoptera % Tanypodinae/TC	% Oligochaeta % Tanypodinae/TC % Noninsect	% Oligochaeta % Odonata % Tanypodinae/TC	% Oligochaeta % Noninsect	% Oligochaeta
Tolerant/Intolerant	NCBI	Dominant Individuals	Tolerant Taxa	HBI	NCBI % Tolerant Individuals HBI
FFG	Scraper Taxa	% Collector % Filterer	% Filterer	Shredder Taxa	Predator Taxa Shredder Taxa
Habit	% Clinger			Sprawler Taxa	

TABLE 5.7

Performance of Macroinvertebrate Metrics by Ecoregion in the State of Georgia

Macroinvertebrate Metric	Average DE (DE for Each Ecoregion/ Total Number of Ecoregions)	Number of Indices Metric Used	% Indices Metric Used ($n = 6$)
North Carolina Biotic Index	0.59	3	50.0
Hilsenhoff's Biotic Index	0.59	2	33.3
%Tanypodina/Total Chironomidae	0.57	3	50.0
% Intolerant Individuals	0.55	2	33.3
% Tolerant Individuals	0.50	0	0.0
Number of Intolerant Taxa	0.50	0	0.0
Beck's Biotic Index	0.49	0	0.0
% Odonata	0.47	2	33.3
% Plecoptera Taxa	0.44	3	50.0
% Oligochaeta	0.43	2	33.3
% Clinger Taxa	0.43	2	33.3
EPT Taxa	0.42	3	50.0
Number of Tolerant Taxa	0.41	0	0.0
% Plecoptera	0.41	2	33.3
% as Hydropsychidae/ Total Trichoptera	0.39	1	16.7
% EPT	0.39	1	16.7
Ephemeroptera Taxa	0.39	0	0.0
% as Hydropsychidae/ Total EPT	0.36	0	0.0
% Noninsect	0.34	1	16.7
Scraper Taxa	0.33	1	16.7
% Scraper	0.33	0	0.0
Clinger Taxa	0.33	1	16.7
% Shredders	0.33	0	0.0
% Chironomidae	0.32	1	16.7
Predator Taxa	0.32	1	16.7
Shredder Taxa	0.32	0	0.0
% Ephemeroptera	0.31	0	0.0
% Cricotopus spp. and Chironomus spp./Total Chironomidae	0.31	0	0.0
% Orth/TC	0.30	0	0.0
Dominant Individuals	0.30	0	0.0

TABLE 5.7 (CONTINUED)
Performance of Macroinvertebrate Metrics by Ecoregion in the State of Georgia

Macroinvertebrate Metric	Average DE (DE for Each Ecoregion/ Total Number of Ecoregions)	Number of Indices Metric Used	% Indices Metric Used ($n = 6$)
% Predators	0.30	1	16.7
% Trichoptera	0.29	1	16.7
Coleoptera Taxa	0.29	0	0.0
Burrower Taxa	0.28	1	16.7
% Coleoptera	0.27	0	0.0
Total Taxa	0.27	0	0.0
Simpson's Diversity Index	0.26	1	16.7
% Filterers	0.26	0	0.0
% Dominant Taxa	0.26	0	0.0
Margalef's Index	0.26	1	16.7
Evenness	0.25	0	0.0
% Isopoda	0.24	1	16.7
Trichoptera Taxa	0.23	0	0.0
Shannon's Index (base e)	0.23	0	0.0
% Tanytarsini/Total Chironomidae	0.22	0	0.0
Sprawler Taxa	0.22	0	0.0
Diptera Taxa	0.22	0	0.0
Chironomidae Taxa	0.21	0	0.0
% Diptera	0.21	0	0.0
Collector Taxa	0.18	0	0.0
Filterer Taxa	0.14	0	0.0
% Baetidae/Total Ephemeroptera	0.13	0	0.0
% Collectors	0.11	0	0.0
Swimmer Taxa	0.11	0	0.0
% Tanytarsini	0.10	0	0.0
% Gastropoda	0.10	0	0.0
Tanytarsini Taxa	0.10	0	0.0
% Amphipoda	0.05	0	0.0
Climber Taxa	0.04	0	0.0

TABLE 5.8
Performance of Macroinvertebrate Metrics by Subecoregion in the State of Georgia

Macroinvertebrate Metric	Average DE (DE for Each Subecoregion/ Total Number of Subecoregions)	Number of Indices Metric Used	% Indices Metric Used ($n = 23$)
Hilsenhoff's Biotic Index	0.58	4	17.4
North Carolina Biotic Index	0.57	7	30.4
% Tolerant Individuals	0.54	5	21.7
% Intolerant Individuals	0.50	5	21.7
Number of Intolerant Taxa	0.49	0	0.0
% Oligochaeta	0.47	8	34.8
Number of Tolerant Taxa	0.45	1	4.3
Beck's Biotic Index	0.45	1	4.3
% Odonata	0.45	4	17.4
% Tanypodini Total Chironomidae	0.44	7	30.4
Shredder Taxa	0.42	5	21.7
Plecoptera Taxa	0.41	7	30.4
% Scrapers	0.41	3	13.0
% Noninsects	0.40	2	8.7
% Chironomidae	0.40	3	13.0
% Filterers	0.40	3	13.0
EPT Taxa	0.39	3	13.0
Ephemeroptera Taxa	0.39	2	8.7
% Plecoptera	0.38	3	13.0
% EPT	0.37	2	8.7
% Clingers	0.37	5	21.7
% Hydrophsychidae/ Total Trichoptera	0.37	4	17.4
% Diptera	0.37	0	0.0
% Shredders	0.36	3	13.0
Scraper Taxa	0.36	8	34.8
% Ephemeroptera	0.35	1	4.3
% Trichoptera	0.35	1	4.3
% Hydropsychidae/Total EPT	0.34	0	0.0
Clinger Taxa	0.34	4	17.4
% Coleoptera	0.34	0	0.0
Coleoptera Taxa	0.33	2	8.7
% Orthcladiane/Total Chironomidae	0.32	1	4.3

TABLE 5.8 (CONTINUED)
Performance of Macroinvertebrate Metrics by Subecoregion in the State of Geórgia

Macroinvertebrate Metric	Average DE (DE for Each Subecoregion/ Total Number of Subecoregions)	Number of Indices Metric Used	% Indices Metric Used ($n = 23$)
Simpson's Diversity Index	0.32	1	4.3
% *Crictopus* spp. and *Chironomus* spp./Total Chironomidae	0.30	2	8.7
Evenness	0.30	0	0.0
Margalef's Index	0.30	1	4.3
Chironomidae Taxa	0.30	2	8.7
% Tanytarsini/Total Chironomidae	0.30	0	0.0
Predator Taxa	0.30	2	8.7
Sprawler Taxa	0.29	4	17.4
% Predators	0.29	1	4.3
Shannon's Index (base e)	0.29	0	0.0
% Dominant	0.29	2	8.7
Number of Dominant Individuals	0.29	2	8.7
Total taxa	0.29	0	0.0
Diptera taxa	0.28	2	8.7
Trichoptera Taxa	0.28	2	8.7
% Tanytarsini	0.26	2	8.7
Collector Taxa	0.25	1	4.3
Burrower Taxa	0.25	2	8.7
% Isopoda	0.25	1	4.3
Tanytarsini Taxa	0.21	1	4.3
Swimmer Taxa	0.21	3	13.0
% Gastropoda	0.20	2	8.7
% Collectors	0.19	2	8.7
Filterer Taxa	0.17	1	4.3
% Amphipoda	0.15	0	0.0
% Baetidae/Total Ephemeroptera	0.13	0	0.0
Climber Taxa	0.09	0	0.0

POTENTIAL RESTRICTIONS ON UTILIZATION OF INDICES

Although it is possible to create a useful set of indices for discrimination between reference conditions and impaired wadeable streams, there are several concerns that may yield restrictions on the interpretation of results that must be taken into account.

In some regions, more than one stream type may exist. Several ecoregions and subecoregions contain both clearwater streams and blackwater streams, dominated by excess tannins, lowered pH, and low dissolved oxygen concentrations. These streams, dominant in some warm humid or subtropical areas may have a unique benthic fauna unlike those of their clearwater counterparts in the same ecoregion or subecoregion. The sandy substrates tend to be dominated by oligochaetes, dipteran larvae, and molluscs that utilize deposited fine particulate organic matter (Meyer 1990), while many snags in the same rivers contain unique and productive assemblages (among the highest of any stream in the country) as well (Benke and Jacobi 1994). Clearwater streams in the same region are likely to be dominated by a different assemblage of macroinvertebrates, containing a greater percentage of Trichoptera and acid-intolerant Chironomids. When analyzing catchments and choosing candidate reference sites, it will be necessary to discriminate between blackwater and clearwater streams. Thus, at both ecoregional and subecoregional levels, the suggested macroinvertebrate indices cannot be a composite of blackwater and clearwater streams in each ecoregion. We suggest that separate reference conditions must be created for blackwater reference streams and for clearwater reference streams in the same ecoregion. To adequately accomplish this task, a greater number of streams in this ecoregion must be sampled and new indices developed.

During the analysis and selection process for candidate reference streams, it may be difficult to determine the influence of tidal flows on many catchments of coastal streams. Some streams, although located many kilometers from the apparent estuary, still contain a proportion of brackish water, freshwater-tolerant marine species, especially polychaetes, some gastropods, and some benthic crustaceans. Although it is possible to identify these organisms, taxonomically, little research has been conducted, which accounts for pollution tolerance and other metrics usually associated with the RBP. As a result, biotic indices derived for these streams systems may underestimate the macroinvertebrate "signature" of a reference condition. The creation of appropriate metrics for tidal streams has yet to be undertaken and we suggest that, ultimately, it will be necessary to create reference conditions for inland streams as well as tidal coastal streams, if both types of streams exist within the same ecoregion or subecoregion.

A number of unusual hydrologic conditions may also confound the ability to create adequate reference conditions. For example, during our studies, the state of Georgia was experiencing a sustained drought during the years of sampling. As a result, many identified candidate reference streams, especially those in the Coastal Plains region and other southeastern Georgia subecoregions were dry for two years or more. Although it has been demonstrated that macroinvertebrates can recolonize a disturbed stream in a relatively short period of time (as little as 14 to 21 days; Gore and Milner 1990), it was difficult to ascertain if these streams had attained an equilibrium or stable recovered community after such an extended period of disturbance or less than optimal flow conditions. As a result, the utility of the macro-invertebrate indices described for these ecoregions must be tempered with a certain amount of uncertainty until such time as these candidate reference streams can be resampled after at least one full year of sustained normal hydrographic flows.

It must be reiterated that every consideration should be made to acquire samples when the catchments being considered are experiencing what are considered to be median precipitation and hydrographic conditions. This will become increasingly important with the prospects of global climate change and the recent research, which indicates that cyclic landscape-level phenomena may have a greater impact on river and stream conditions than previously imagined. As a result, it may be necessary to create "sets" of reference conditions to be applied to the appropriate cyclic condition and the proposed consequences of global climate change. Gore et al. (2008) have demonstrated that the effects of a cyclic temperature change, known as the Atlantic multidecadal oscillation (AMO), sufficiently alter precipitation patterns on an approximate 30- to 40-year cycle and there can be significantly different wet hydrographic periods and dry hydrographic periods throughout the southeastern United States; with the last dry hydrographic period between 1940 and 1969 and the latest wet hydrographic period between 1970 and 1999. Further, Gore et al. (2008) have predicted that the availability of habitat for various fish and invertebrate species is so different during these two cycles that community composition and dynamics will be dramatically different. As a result, creation of a suite of reference conditions during a dry hydrographic period will not be an appropriate point of comparison during a wet hydrographic period. Similar changes in water quality (Monteith et al. 2000) and macroinvertebrate community composition (Bradley and Ormerod 2001) have been observed as a result of the impacts of the North Atlantic oscillation (Hurrell and van Loon 1997) in the United Kingdom. However, these impacts may influence river systems across Europe and into the Middle East (Cullen and de Menocal 2000). Similar observations of Pacific oscillations suggest that rainfall and resulting hydrographic changes will yield the same long-term changes in community structure in regions affected by Pacific weather influences (see, for example, Biondi et al. 2001 and Salinger et al. 2001). Since these cyclic events occur over a period of decades, institutional memory and records should be sufficient to enable creation of a new set of reference conditions to account for new meteorological conditions over the coming set of decades.

REFERENCES

Beck, W.M., Jr. 1965. The streams of Florida. *Bulletin of the Florida State Museum* 10: 81–126.

Benke, A.C., and D.I. Jacobi. 1994. Production dynamics and resource utilization of snag-dwelling mayflies in a blackwater river. *Ecology* 75: 1219–1232.

Biondi, F., A. Gershunov, and D.R. Cayan. 2001. North Pacific decadal climate variability since 1661. *Journal of Climate* 14: 5–10.

Bradley, D.C., and S.J. Ormerod. 2001. Community persistence among stream invertebrates tracks the North Atlantic Oscillation. *Journal of Animal Ecology* 70: 987–996.

Cullen, H.M., and P.B. de Menocal. 2000. North Atlantic influences on Tigris-Euphrates streamflow. *International Journal of Climatology* 20: 853–863.

Glantz, S.A. 1992. Alternatives to analysis of variance and the *t* test based on ranks. In *Primer of Biostatistics*, 3rd ed., ed. S. Glantz, 320–343. New York: McGraw-Hill.

Gibson, G.R., M.T. Barbour, J.B. Stribling, J. Gerritsen, and J.R. Karr. 1996. *Biological Criteria: Technical Guidance for Streams and Small Rivers* (rev. ed.). EPA 822-B-96-001. Washington, DC: U.S. Environmental Protection Agency, Office of Science and Technology.

Gore, J.A., A.F. Casper, and M.H. Kelly. 2008. *The Atlantic Multidecadal Oscillation (AMO) as a surrogate for climate change in the southeastern United States*. Paper presented at North American Benthological Society, Salt Lake City, UT.

Gore, J.A., and A.M. Milner. 1990. Island biogeographic theory: Can it be used to predict lotic recovery rates? *Environmental Management* 14: 737–753.

Gore, J.A., J.R. Olson, D.L. Hughes, and P.M. Brossett. 2005. *Reference Conditions for Wadeable Streams in Georgia with a Multimetric Index for the Bioassessment and Discrimination of Reference and Impaired Streams*. Atlanta: Georgia Department of Natural Resources.

Hurrell, J.W., and J. van Loon. 1997. Decadal variations in climate associated with the North Atlantic Oscillation. *Climate Change* 36: 301–326.

Jessup, B.K., and J. Gerritsen. 2000. *Development of Multimetric Index for Biological Assessment of Idaho Streams Using Benthic Macroinvertebrates*. Prepared for the Idaho Department of Environmental Quality. Owings Mills, MD: Tetra Tech, Inc.

Lenat, D.R. 1993. A biotic index for the Southeastern United States: Derivation and list of tolerance values, with criteria for assigning water-quality ratings. *Journal of the North American Benthological Society* 12: 279–290.

Merritt, R.W., and K.W. Cummins, eds. 1996. *An Introduction to the Aquatic Insects of North America*. Dubuque, IA: Kendall/Hunt.

Meyer, J.L. 1990. A blackwater perspective on riverine ecosystems. *BioScience* 40: 643–651.

Minshall, G.W. 1988. Stream ecosystem theory: A global perspective. *Journal of the North American Benthological Society* 7: 263–268.

Mississippi Department of Environmental Quality (MDEQ). 2003. *Development and Application of the Mississippi-Benthic Index of Stream Quality (M-BISQ)*. Jackson: Mississippi Department of Environmental Quality, Office of Pollution Control.

Monteith, D.T., C.D. Evans, and B. Reynolds. 2000. Are temporal variations in the nitrate content of UK upland freshwaters linked to the North Atlantic Oscillation? *Hydrological Processes* 14: 1745–1749.

Rosenberg, D.M., and V.H. Resh. 1996. Use of aquatic insects in biomonitoring. In *An Introduction to the Aquatic Insects of North America*, 3rd ed., eds. R.M. Merritt and K.W. Cummins, 87–98. Dubuque, IA: Kendall/Hunt Publishing.

Salinger, M.J., J.A. Renwick, and A.B. Mullan. 2001. Interdecadal Pacific Oscillation and South Pacific climate. *International Journal of Climatology* 21: 1705–1721.

Stribling, J.B., B.K. Jessup, J.S. White, D. Boward, and M. Hurd. 1998. *Development of a Benthic Index of Biotic Integrity for Maryland Streams*. CBWP-EA-98-3. Annapolis: Maryland Department of Natural Resources, Monitoring and Non-Tidal Assessment Division.

6 A Numerical Index of Stream Health

*Amanda M. Herrit, Duncan L. Hughes,
James A. Gore, and Michele P. Brossett*

CONTENTS

As part of the Clean Water Act (CWA) (Section 101(a)), it is the obligation of each state to monitor and assess the chemical, physical, and biological conditions of streams within its boundaries. Therefore, states are required to consider the "biological integrity" of their waters when developing stream monitoring procedures (Berry and Dennison 2000). Biotic indices are accepted by the Environmental Protection Agency (EPA) as a method for assessing the biological health or condition of wadeable streams (Barbour et al. 1999).

States must also determine water quality standards for all water bodies as required by CWA Section 305(b). Water quality standards establish designated use and criteria for each water body, which must be maintained for all waters within each state (Berry and Dennison 2000). State agencies must first define water quality standards and then determine a method of monitoring these standards. In some states, biological indices have been used to assess and monitor streams in order to maintain water quality standards set by states throughout the United States (Barbour 1997).

Beginning in the 1970s with the CWA, biological monitoring has developed into a widely used tool for tracking the condition of water resources. In the United States, the chemical condition of water resources was the primary consideration in monitoring and remediating processes before the 1970s (Berry and Dennison 2000). During the last 20 years, the United States has made great improvements eliminating point source pollution and, as a result, chemical contamination has been greatly reduced. More recently, the major impairment concern for surface waters is nonpoint source pollution (Barbour 1997). Biological assessment has been found to be an equally effective tool for assessing both point and nonpoint source pollution (Karr 1991).

For water bodies that have been described as impaired, states must develop a plan for returning that water body to an unimpaired status. Important regulatory controls, intended to accomplish this task, are total maximum daily loads (TMDLs) of target nonpoint source contaminants, ranging from metals and nutrients to suspended sediment. According to CWA Section 303(d), state regulatory agencies must establish TMDLs for each water body that has not attained water quality standards after imposing technology-based controls (Barbour et al. 1999). Biological assessments of the structure and function of lotic communities can determine whether water quality standards have been achieved and if TMDLs are required for a specific water body.

A broadly applicable indicator for use in biological assessment is the Index of Biotic Integrity (IBI) (Karr 1981). The IBI approach was developed to identify levels of stream impairment using the fish assemblage as a biological indicator. Using the IBI as a model, many biomonitoring programs have expanded to incorporate several types of multimetric indices using fish, macroinvertebrates, and periphyton assemblage-level data. It has been shown that using multiple assemblages from various trophic levels can provide assessments of a broader array of stressors causing stream impairment (Karr 1991).

By monitoring biological indicators, such as the benthic macroinvertebrate community, monitoring agencies can describe a given stream or river condition. Macroinvertebrates are considered excellent indicators because they are relatively sedentary and, thus, can be used to assess long-term change and cumulative effects in a specific location. Depending upon the number of sampling locations and monitoring network design, biological indicators can be used in broader-scale assessments, such as catchments, subecoregions, or ecoregions.

According to Murtaugh (1996), an indicator is considered effective if it is sensitive to stressors or other specific factors being considered. When investigating stream conditions, macroinvertebrate community assemblages can provide researchers with a description of the stream's condition (Resh 1995). Stream community structures are altered by human disturbance and can be used to identify the type and level of disturbance encountered. Using ecological descriptors such as tolerance/intolerance values, macroinvertebrate assemblages can describe the impairment level of a stream relative to the chosen reference condition (Barbour et al. 1999).

The Rapid Bioassessment Protocol (RBP) was developed as a cost-effective and time-efficient procedure for assessing wadeable streams (Barbour et al. 1999). In its most complete, but rarely applied, form, the RBP uses fish, benthic macroinvertebrate, and periphyton assemblage-level data to develop a multimetric index, which is used as the indicator of stream impairment. Metrics are used to quantify different attributes of the stream biota (Jessup and Gerritsen 2000). The choice of final metrics ultimately used in an index is based, in part, on their relationship to ecoregional characteristics and response to stressors (Barbour et al. 1999).

Several methods have been used to classify streams according to their abiotic characteristics. Ricker (1934) developed a stream classification for streams in Ontario, Canada, based upon the size of the stream, substrate material, the diversity and abundance of the biota, and the physical and chemical characteristics of the water body. Ricker developed this system to group streams according to their similar abiotic characteristics. This technique was used to investigate streams with similar physical and chemical properties, therefore defining each stream's biological characteristics without biased abiotic information.

Multimetric indices are used to describe the ecological characteristics and to detect threats to the biological integrity of a stream (Rankin 1995). As described in Chapter 2, metrics from richness, composition, tolerance/intolerance, and habit/trophic biological categories are evaluated to determine their ability to detect differences in reference and impaired conditions. Streams are grouped according to their physical and chemical characteristics and are compared within groups. Between six to eight metrics are usually chosen for an index and assigned a quantitative index score for each stream. Based on variation from the least impaired sites, the index score describes each stream relative to their level of impairment. Once a quantitative rating is assigned, the index score can be described by a qualitative rating.

Using quantitative index scores to describe streams within groups, narrative ratings describe stream characteristics qualitatively. Narrative ratings typically group streams into the qualitative categories of *good, fair,* and *poor.* Each stream is evaluated based on its potential to achieve the least impaired condition within each group. Qualitative measures of stream condition can be used to determine regulatory and monitoring needs of each stream. Using narrative biological criteria, monitoring agencies can determine action plans for stream conservation and restoration (Karr and Chu 1999).

In 1977, Hilsenhoff introduced his biotic index based on organic and nutrient tolerance/intolerance levels of arthropods. Using one phylum, Hilsenhoff was able to simplify the bioassessment process. Hilsenhoff's biotic index was based on a 100 individual sample in which each species or genus of arthropod was assigned a tolerance or intolerance level. Once all individuals from each sample were identified, the tolerance/intolerance values were averaged together giving each stream a biotic index score (Hilsenhoff 1987).

Originally, the CWA standard for adequate biological support was termed "fishable–swimmable" but this standard has evolved into a more functional "aquatic life use" designation (Berry and Dennison 2000). Multimetric indices can also be used to determine aquatic life use designations; an EPA requirement for nonpoint source management. The Vermont Department of Environmental Conservation has employed benthic and fish data, by means of multimetric indices, to determine numeric biological criteria. Numeric criteria are applied and are used to evaluate each water body according to aquatic life use designations. Being the quantitative equivalent of narrative biological criteria, numeric biological criteria can also be used to assess water quality standards (Vermont Department of Environmental Conservation 2004)

Several states have developed narrative rating systems to describe numeric biological criteria. In Ohio, the Qualitative Habitat Evaluation Index (QHEI) was developed to determine the aquatic life potential of each water body. Each water body is assigned an aquatic use level, which could be applied to aquatic life use designations. The purpose of this system was to describe the physical, chemical, and biological properties of a water system, and therefore protect all facets of this system (Rankin 1989). Narrative criteria are used to describe the water body's condition or current state, which is based on quantitative data. The QHEI has two main categories of aquatic life uses: warm water habitat (WWH) and exceptional warm water habitat (EWH). The WWH is described as the typical habitat condition of rivers and streams in Ohio. The EWH is an aquatic habitat that is exceptional for its fauna and quality of habitat. Narrative criteria of exceptional (EWH), good (WWH), fair, poor, or very poor are assigned to each stream or river (Rankin 1989).

The Benthic Index of Stream Integrity (BISI), developed in Rockdale County, Georgia, assigns each stream with a quantitative rating. Using a percentile method, the index score is described by a qualitative rating. Streams with an index score above the 25th percentile are equally divided into *good* and *very good* narrative ratings. Streams rated below the 25th percentile are divided into three groups: *fair, poor,* and *very poor.* Narrative ratings are used to describe biological characteristics that are found in each stream category (Tetra Tech 2001b).

With the use of multimetric indices, chemical analysis, and physical habitat assessment, stream assessment methods have been developed to identify the level of stream impairment. Once stream assessment is completed, this information can be used to determine regulatory and monitoring procedures for the study area. The evaluation of stream conditions is an important method for managing water resources (Barbour et al. 1999).

Ultimately, the objective is to produce a rating system of stream status that can be used to determine the impairment state of streams within a geographical region, ecoregion, or subecoregion. By the use of multimetric indices, the rating system designates a numeric value for each stream. This numeric value can be used to determine regulatory action for each stream within an ecoregion. The stream rating system we describe incorporates benthic macroinvertebrate, chemical, and physical habitat data to produce a robust assessment tool. This level of evaluation has been the most commonly applied. Although few regulatory agencies have attempted the endeavor, Barbour et al. (1999) suggest that employing multiple trophic levels (periphyton, macroinvertebrates, and fish) will increase the sensitivity of the RBP assessment. The assessment system will supplement previous assessment methods from other states (MDEQ 2003) by providing a method for not only assessing streams but also regulating compliance with water quality regulations for streams in ecoregions or subecoregions.

METHODS

The complete field sampling and laboratory methods are found in Chapter 3. These data were employed to select the candidate reference sites (as described in Chapter 4), to evaluate and select impaired sites and elucidate the reference condition (Chapter 5) and, in combination, to derive the numerical index of stream health described in this chapter.

Based on the raw macroinvertebrate data, a multimetric analysis calculated by the Ecological Data Application System (EDAS) (Microsoft Access 2000; Tetra Tech 2001a) is used to assess stream condition. Metrics are selected from the following categories of biological information: richness, composition, tolerance/intolerance, and habit/trophic measures, so that each category is represented when possible. Metrics are grouped into candidate indices for each ecoregion (Gore et al. 2005).

The following protocol (as described in Chapter 5) is used for developing ecoregion-based multimetric indices and subsequent classification systems. All data are entered, quality checked, and metrics calculated using EDAS (Microsoft Access 2000; Tetra Tech 2001a) or other suitable spreadsheet programs. Various statistical software packages (we have used Statistica [Statsoft 2000] successfully) can be used to run Pearson's correlation and box-and-whisker plots (MDEQ 2003).

Pearson's *r*-correlation is used to determine redundancy among metrics. Box-and-whisker plots are used to demonstrate the ability of different indices to discriminate between stream conditions. Desirable indices will demonstrate a complete separation in box-and-whisker plots (i.e., no overlap of interquartile ranges) between reference and impaired conditions.

When two metrics are calculated to have Pearson's *r*-correlation values of greater than 0.90 or less than −0.90, one is automatically eliminated from candidate metrics because of redundancy with other metrics. Metrics with a Pearson's *r*-value of 0.80 to 0.90 or −0.80 to −0.90 can be considered as candidate metrics if the relationship is not similar to correlations between that value and other metrics. If candidate metrics have a parallel linear relationship, their relationship should be considered to be codependent and thus the information provided by that metric does not provide additional discrimination and one of those two metrics should be eliminated as a reference metric. Once undesirable metrics are eliminated, final candidate metric scores are standardized to a 100-point scale (Mississippi Department of Environmental Quality [MDEQ] 2003).

From final candidate metric scores, several candidate indices are selected, each including four to seven metrics. Metrics are selected to represent each structural and behavioral category, to discriminate between reference and stressed conditions, and to produce unique information for each index. Each index is compared using the discrimination efficiency (DE) and box-and-whisker plots. The discrimination efficiency and box-and-whisker plots reveal whether each candidate index discriminates between reference and impaired conditions. The index with the greatest discrimination ability should be selected (MDEQ 2003). Selection of the final indices should consider the metric selection criteria and chemical and physical data. Any other selection criteria are based on best professional judgment. The ideal index has a box-and-whisker plot with good discrimination efficiency, has little or no overlap between reference and impaired conditions, allows detection of stream impairment, and ranks relative severity of impairment.

The final analytical product is the formulation of a numeric rating system for the stream systems selected, in the context of ecoregional or subecoregional differences. Each stream classification level has a multimetric index that designates a specific impairment condition. Using multimetric indices and abiotic data, streams are grouped into levels of impairment and are each given a numeric assessment of 1 to 5 (1 = *very good*; 2 = *good*; 3 = *fair*; 4 = *poor*; 5 = *very poor*).

THE NUMERIC RANKING SYSTEM

To develop an effective ranking system, it is necessary to sample a sufficient number of candidate reference and impaired sites from each ecoregion and subecoregion to assure that calculation of discrimination efficiencies are defensible indicators of stream health. Composite indices of benthic macroinvertebrate response must be developed for each ecoregion and subecoregion. This may require several index periods of sampling over several years. For example, in the state of Georgia, the development of a numerical index required physical, chemical, and biological data collected from a total of 111 candidate reference sites and 184 impaired sites.

TABLE 6.1
Stream Rating Based on Numeric Ranking

Numeric Ranking	Stream Health Rating	Management Decision
1 or 2	A	Continue periodic monitoring to detect change in baseline reference condition
3	B	Frequent monitoring critical to detect change in ecological status, lower range especially
4 or 5	C	Frequent monitoring necessary to determine remediation needs and if remediation has been successful

During the development of the numerical index, each stream receives an index score. The index score is the average of all standardized metric values used in the index. Each stream also is ranked, described, and rated. A stream receives a ranking between 1 and 5, which corresponds with a narrative description of very good, good, fair, poor, and very poor. In the state of Georgia, this classification of the stream's "health" rating combined the two top categories of very good and good for an *A* rating, fair for a *B* rating, and poor and very poor for a *C* rating (Table 6.1).

It should not be surprising that subecoregion-level indices have higher discrimination efficiencies than ecoregion-level indices. Subecoregions with smaller catchment areas, being more similar in geomorphology and chemical signature, will also tend to have higher discrimination efficiencies than subecoregions with larger catchment areas. As a result, very different indices of benthic macroinvertebrate community structure and composition will separate reference (minimally impaired) and impaired streams. For example, in Georgia, indices for ecoregions in the piedmont and mountain areas tended to contain metrics from all functional and structural categories, especially richness, whereas the larger Southeastern Plains ecoregion and the Southern Coastal Plains ecoregion had indices developed primarily from metrics in the composition category and rarely from richness category. As in Chapter 5, we have appended examples of the numerical rating system and discrimination efficiencies from the state of Georgia (Appendix F).

DISCUSSION AND RECOMMENDATIONS

The numerical rating system is based upon a percentile assignment. In each ecoregion, streams with index scores above or equal to the 95th percentile are given the numeric value of 1 or rated as having a *very good* stream condition. Index scores that fall below the 95th percentile yet are equal to or above the 75th percentile receive a numeric value of 2 or are rated as having a *good* stream condition. Streams with index scores below the 75th percentile and above the 25th percentile receive a *fair* stream rating. The poor rating is assigned to streams with index scores equal to or below the 25th percentile yet above the 5th percentile. For streams equal to or below the 5th percentile, the *very poor* rating is assigned. These ratings are arbitrary; however, they offer a system of ranking for the purpose of prioritizing management activities.

For management purposes, narrative categories also can be developed. Categories 1 and 2 of the numeric rating system are combined to create a group of least impaired sites. We recommend that least impaired sites require periodic monitoring to evaluate change over time relative to the reference condition. Indeed, the reference streams (those in Category 1) should be reevaluated so that the "reference condition" (see Chapter 5 and Gore et al. 2005) can be modified as new information and samples are added to the analytical system. Category 3 encompasses the majority of streams, which vary widely in their numeric range. Category 3 streams are impaired streams that require frequent monitoring to determine condition change over time. Streams rated as *poor* and *very poor* are combined into one group for severely impaired sites. Severely impaired sites require frequent monitoring to determine restoration needs and success of restoration attempts.

It must be remembered that the numerical ranking system described in this chapter as well as the "reference condition" for wadeable streams are not static in their application. The process of determining the health of streams, using rapid bioassessment, is a dynamic process. The system of classifying reference conditions and the creation of the numerical ranking system is based upon information contained within a continuously growing data bank. As a monitoring system for the entire state is implemented, new physicochemical and biological data will become available for both reference and impaired streams. When these data are entered into the analytical system, it will be possible (and it should be imperative) to reevaluate the reference condition for each ecoregion and subecoregion. Subsequent to the creation of a new set of reference criteria and metric indices, a new numerical classification system can also be developed. Thus, the classification of "health" becomes a continually evolving process as reference conditions are more closely defined and as effective management continues to improve the condition of impaired streams; the ultimate goal being the detection of no impaired streams in each ecoregion. We recommend that the reevaluation of each ecoregion and subecoregion take place at least every five years, as new stream data become available.

The multimetric rating system can be used to guide nonpoint source regulations for Georgia's streams. Using the macroinvertebrate community as a biomonitor, the multimetric rating system uses fundamental characteristics of the stream community to determine the overall health or condition of that community. By using macroinvertebrates assemblages as a reference for stream condition, the multimetric rating system can be used to describe the condition for the entire stream community and therefore determine overall stream condition or health.

The multimetric method is a valid assessor of biological systems because of its ability to describe interactions among several levels of the lotic community, to discriminate between natural and impaired conditions, and to detect a range of stream conditions (Barbour et al. 1996). The multimetric rating system determines levels of impairment within the subecoregion or ecoregion level relative to the least impaired condition in the same region. By developing indices based on multiple metric data, the multimetric rating system can be used as a tool for making sound water management decisions.

When developing the multimetric rating system and prioritization list, each stream should be described according to the macroinvertebrate data collected and not according to the stream's original classification when creating the metric index. As a result, all streams are rated independent of their reference or impaired classification. For instance, an impaired stream may rate as a *B* or *good* stream, and vice versa, a reference stream may rate as a *C* or *poor/very poor* stream.

Within each ecoregion, the discrimination efficiencies (DEs) for each individual subecoregion tended to be much higher than the ecoregion-level DE. In Georgia, smaller ecoregions containing fewer subecoregions were found to have higher DE than ecoregions containing geographically widespread subecoregions. This reveals that subecoregion-level indices more easily discriminate between reference and impaired conditions than ecoregion-level indices. As a result, multimetric indices are most effective when used to describe stream condition on a more localized scale such as the subecoregion level.

The multimetric rating system has many applications for water resource management. The multimetric rating system is a simple to use, relatively inexpensive, and extremely versatile water management tool. By grouping streams into three stream condition categories, the multimetric rating system is a simple method for determining biological and ecological conditions within a stream system, as well as overall water quality. The multimetric rating system requires only limited technical knowledge, which allows ease of use for water resource managers. This system allows states to begin the process of classifying streams, as prescribed by Section 319 of the Clean Water Act, and to prioritize their efforts in restoring impaired streams and maintaining the health of the least impaired streams in the same ecoregions.

REFERENCES

Barbour, M.T. 1997. The re-invention of biological assessment in the U. S. *Human and Ecological Risk Assessment* 3: 933–940.

Barbour, M.T., J. Gerritsen, G.E. Griffith, R. Frydenborg, E. McCarron, J.S. White, and M.L. Bastian. 1996. A framework for biological criteria for Florida streams using benthic macroinvertebrates. *Journal of the North American Benthological Society* 15: 185–211.

Barbour, M. T., J. Gerritsen, B.D. Snyder, and J.B. Stribling. 1999. *Rapid Bioassessment Protocols for Use in Streams and Wadeable Rivers: Periphyton, Benthic Macroinvertebrates and Fish*, 2nd ed. EPA 841-B-99-002. Washington, DC: U.S. Environmental Protection Agency, Office of Water.

Berry, J.F., and M.S. Dennison. 2000. *The Environmental Law and Compliance Handbook*. New York: McGraw-Hill.

Gore, J.A., J.R. Olson, D.L. Hughes, and P.M. Brossett. 2005. *Reference Conditions for Wadeable Streams in Georgia with a Multimetric Index for the Bioassessment and Discrimination of Reference and Impaired Streams*. Atlanta: Georgia Department of Natural Resources.

Hilsenhoff, W.L. 1987. An improved biotic index of organic stream pollution. *The Great Lakes Entomologist* 20: 31–39.

Jessup, B.K., and J. Gerritsen. 2000. *Development of Multimetric Index for Biological Assessment of Idaho Streams Using Benthic Macroinvertebrates.* Prepared for the Idaho Department of Environmental Quality. Owings Mills, MD: Tetra Tech, Inc.

Karr, J.R. 1981. Assessment of biotic integrity using fish communities. *Fisheries* 6: 21–27.

Karr, J.R. 1991. Biological integrity: A long-neglected aspect of water resource management. *Ecological Applications* 1: 66–84.

Karr, J.R., and E.W. Chu. 1999. *Restoring Life in Running Waters: Better Biological Monitoring.* Washington, DC: Island Press.

Mississippi Department of Environmental Quality (MDEQ). 2003. *Development and Application of the Mississippi-Benthic Index of Stream Quality (M-BISQ).* Jackson: Mississippi Department of Environmental Quality, Office of Pollution Control.

Murtaugh, P.A. 1996. The statistical evaluation of ecological indicators. *Ecological Applications* 6: 132–139.

Rankin, E.T. 1989. *The Qualitative Habitat Evaluation Index (QHEI): Rationale, Methods, and Application.* Columbus: Ohio Environmental Protection Agency, Ecological Assessment Section, Division of Water Quality, Planning, and Assessment.

Rankin, E.T. 1995. Habitat indices in water resource quality assessment. In *Biological Assessment and Criteria: Tools for Water Resource Planning and Decision Making*, eds. W.S. Davis and T.P. Simon, 181–208. New York: Lewis Publishers.

Resh, V.H. 1995. Freshwater benthic macroinvertebrates and rapid assessment procedures for water quality monitoring in developing and newly industrialized countries. In *Biological Assessment and Criteria: Tools for Water Resource Planning and Decision Making,* eds. W.S. Davis and T. P. Simon, 167–177. New York: Lewis Publishers.

Ricker, W.E. 1934. *An Ecological Classification of Certain Ontario Streams* (University of Toronto Press, Publications of the Ontario Fisheries Research Laboratory, No. 49, University of Toronto Studies, Biological Series No. 37). Toronto: University of Toronto Press.

Tetra Tech, Inc. 2001a. Ecological Data Application System (EDAS), version 3.3.2k [computer software]. Owings Mills, MD: Tetra Tech, Inc.

Tetra Tech, Inc. 2001b. *Watershed Characterization Report, Rockdale County Report.* Rockdale, GA.

Vermont Department of Environmental Conservation. 2004. *Biocriteria for Fish and Macroinvertebrate Assemblages in Vermont Wadeable Streams and Rivers: Development Phase.* Burlington, VT: Vermont Department of Environmental Conservation, Water Quality Division, Biomonitoring and Aquatic Studies Session.

7 The Effect of Sample Size on Rapid Bioassessment Scores

*Uttam K. Rai, James A. Gore,
Duncan L. Hughes, and Michele P. Brossett*

CONTENTS

INTRODUCTION

Rapid Bioassessment Protocol (RBP) applies shortcut techniques in its biomonitoring procedures and these have been achieved, along with other things, by limiting the number of benthic invertebrates selected for processing. The original benthic macroinvertebrate protocols generally required the collection of 100 organisms

95

(e.g., Hilsenhoff 1987). Most states have endorsed this sample size for its ease and speed of processing. The organisms are identified, and the pollution tolerance values (assigned by best professional judgments) of the most abundant taxonomic groups are scored on a scale ranging from 1 to 10. A high score indicates poor stream quality from which the macroinvertebrate were collected. Alternately, the second approach (the RBP) requires an initial characterization (including benthic macro-invertebrate scores) of the reference conditions from similar water bodies that have acceptable water quality (Barbour et al. 1999). The current RBP calls for 200 organisms in order to estimate the health of the water body. About 65% of state regulatory agencies subsample 200 or fewer individuals (Carter and Resh 2001).

The size of the subsample (number of organisms sorted, identified, and cataloged) is an essential problem as it is impossible to completely census a taxonomic assemblage or entire community. Instead, estimates that describe some portion of the real taxonomic richness of an assemblage are relied upon. Hence, the optimum subsample size has been a matter of much debate. The recommended fixed count of 200 organisms is assumed to adequately represent the benthic community of the stream from which it was sampled (Barbour et al. 1999). However, obtaining an adequate, representative sample of ecological communities to make compositional comparisons is a challenge (Cao et al. 2002). How well a sample represents the community normally cannot be measured directly because the taxonomic composition and relative abundance in a community are unknown. The species-area relationship is a well-known ecological pattern, which generally shows that a larger area (that is, a larger subsample) will harbor greater diversity (MacArthur and Wilson 1967). It is still not clear how well a subsample size of 200 organisms captures the taxonomic composition and relative abundance at the sampling site or of the communities being surveyed. As a result, biases and dependency on sample sizes can be introduced into community comparisons. It has been argued that the RBPs give a biased measurement of taxonomic richness because of the density factor (Courtemanch 1996). The community density factor analysis supports the notion that the number of taxa encountered increases as a function of the number of individuals in the sample and the area sampled. Sovell and Vondracek (1999) demonstrated that increasing subsample sizes changed the taxonomic richness metrics, however, the remaining RBPs metrics were not affected. Similarly, Vinson and Hawkins (1996) suggested using greater than 300 organisms to obtain more accurate inferences for richness. Studies by Cao et al. (1998, 2002) demonstrate that the estimation of relative differences in taxonomic richness among sites or communities can be strongly dependent upon the sample sizes and that small samples tend to underestimate the differences. Growns et al. (1997) also support this view, the argument being that small subsample sizes express estimates of the richness of abundant taxa, whereas they often fail to account for taxa that are rare or less abundant. A taxon is determined to be rare if its relative abundance in a community is small (say, less than one individual per square meter). However, rare taxa may be very important components of community integrity because of their tolerance to potential stressors, specialized niche, and functional redundancy.

Several studies have reported comparisons of the size of subsample and how it relates to biological metrics, but few of these studies have been performed on streams inside the United States. Except for Sovell and Vondracek (1999), these studies were performed primarily on lake (Somers et al. 1998) or stream assemblages in Australia (Growns et al. 1995, 1997; Metzelling and Miller 2001). Sovell and Vondracek (1999) used single habitat samples (riffles) for their study. However, Ostermiller and Hawkins (2004) have recently investigated stream samples from Oregon and Washington for errors associated with sample sizes for the River Invertebrate Prediction and Classification System (RIVPACS) and recommended 350 or more individuals. In this chapter we discuss the evaluation of the RBP metric scores using multihabitats (Rai 2006).

The hypothesis was that the analysis of different subsamples of 100, 200, and 300 organisms would produce different metric values.

SITE SELECTION

Thirty stream sites were chosen for the study. These were all third-order or smaller streams. All study sites were part of a larger set of stream sites that were previously selected using land use data and geographic information systems during the characterization of reference stream conditions for Georgia (see Chapters 1 to 3 and Gore et al. 2004). Sites were selected on a longitudinal transect across the ecoregions of Georgia to capture the variability of stream gradients and covered 5 ecoregions and 17 subecoregions of the state. The distribution of these sites across the ecoregions and subecoregions of Georgia are depicted in Figure 7.1.

Following the field and laboratory collecting and sampling protocols of the RBP, as described earlier in Chapter 3, we altered the subsampling routine to allow us to

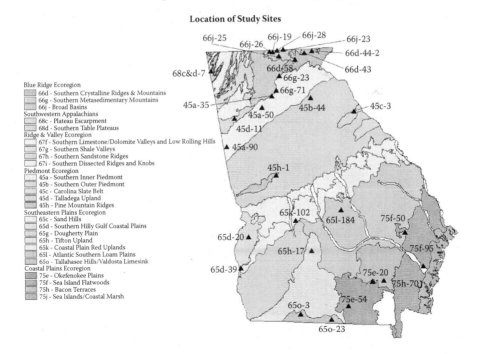

FIGURE 7.1 Distribution of sample sites used to analyze the effect of subsample size in the RBP process.

examine different lots of 100, 200, and 300 individuals from the same composite sample.

In the laboratory, each sample was transferred from the polypropylene bottles into the sieve bucket and thoroughly rinsed with tap water. The sample was then spread evenly across a standardized gridded pan (Caton 1991). The pan contained 30 clearly marked squares, and therefore, would divide the sample into 30 equal portions. Squares to be sorted were randomly chosen. All macroinvertebrates encountered in each square were sorted and collected in a glass vial. Succeeding squares were sorted, where necessary, until 100 organisms were obtained. This comprised the first subsample for the study. Sorting was continued to obtain another 100 organisms and collected in a separate vial. The combination of these organisms with the first subsample made up the second subsample (200 organisms). Similarly, another 100 organisms were sorted and the combination of this to the second subsample made up the third subsample (300 organisms). All specimens were identified to the lowest possible taxon.

METRIC SELECTION

The metrics analyzed for each subecoregion in this study had been predetermined as part of the overall Georgia Ecoregions Project (Gore et al. 2005), using a multimetric (a total of 59 metrics grouped into 5 categories) approach to assimilate biological data with various functional abilities into a single index to gauge the health of a stream. For each ecoregion, the final indices comprised of five to seven individual metrics (with at least one metric chosen, wherever possible, from each of the categories) that best distinguished the reference and impaired streams for that particular subecoregion (see Chapter 5).

BOOTSTRAP RESAMPLING

Since the metric scores (as a combination of nonlinear metrics) could not be demonstrated to be normally distributed, the bootstrap resampling method was chosen to approximate the distribution of possible values associated with each subsample (Efron and Tibshirani 1993).

Using random sampling with replacement for 100 individuals, 25 such samplings were performed to ensure a stable and representative distribution of metric values. Next, the process was repeated to select 200 organisms 25 times. The same was done to select 300 organisms. Thus, a single site had 75 total bootstrap samples, 25 each for 100, 200, and 300 organisms.

ANALYSIS

After generating the 75 replicates at each specified subsample size (100, 200, 300), raw metric values were calculated using the Ecological Data Application System (EDAS), Version 3.3.2k, program (Gore et al. [2005] describe the standardization process).

Final indices obtained from 25 replicates of each subsample (i.e., 100, 200, and 300 organisms) were plotted in box-and-whisker graphs to evaluate how the indices were distributed on a scale scoring from 0 to 100. Variability for box and whiskers was set at the 25th (lower) and the 75th (upper) percentiles to keep the analysis consistent with the method used during metric development (Gore et al. 2005).

The multiple-range test (Steel and Torrie 1960) was used to compare the mean macroinvertebrate index across the range of subsample sizes, allowing for simultaneous comparisons of more than two means.

THE EFFECT OF SUBSAMPLE SIZE ON METRIC SCORES

With few exceptions, increasing subsample size resulted in significantly different (usually higher) composite metric scores. These results are summarized in Table 7.1. Some typical examples of the analyses from various ecoregions provided in Table 7.1 follow.

SUBECOREGION 45A–SOUTHERN INNER PIEDMONT

45a-90

Richness (EPT taxa) increased with increasing subsample size while the remaining metric indices did not display notable changes (Figure 7.2 and Table 7.2). Index variability was greater in the 200-organism subsample. Interquartile variability overlapped between subsamples of 200 and 300 organisms, while scores from subsamples of 100 individuals were substantially lower (no overlap).

SUBECOREGION 65D–SOUTHERN HILLY GULF COASTAL PLAIN

65d-39

Richness index (as Plecoptera and Trichoptera taxa) increased with increasing subsample size. Functional feeding group (FFG) index values (as percent filterer) decreased when subsample size was increased. The remaining metric indices did not display any trends over the range of subsample sizes (Figure 7.3 and Table 7.3). Index variability declined with increasing subsample size. Interquartile variability overlapped between 100- and 200-organism subsamples, while that of the 300-organism subsample did not and was substantially higher.

SUBECOREGION 65K–COASTAL PLAIN RED UPLANDS

65k-102

Only the FFG indices (as scraper taxa and percent shredders) changed slightly when larger subsamples were used. The remaining metric indices did not show any clear trends. Gastropoda were absent from all subsamples (Figure 7.4 and Table 7.4). Interquartile variability and median value declined with larger subsample size. Interquartile variability overlapped among all three subsamples.

SUBECOREGION 66D–SOUTHERN CRYSTALLINE RIDGES AND MOUNTAINS

66d-43

Richness (as Diptera taxa) and habit (as clinger taxa) increased in larger subsamples, while the remaining metric indices did not display any trends (Figure 7.5 and Table 7.5).

TABLE 7.1

Mean Metric Scores from Sample Streams after Bootstrap Resampling 25 Times

Stream and Subecoregion	Mean Standardized Metric Score (100 Individuals)	Mean Standardized Metric Score (200 Individuals)	Mean Standardized Metric Score (300 Individuals)
	Piedmont Ecoregion		
45a-35 Southern Inner Piedmont	54.69	57.06	56.94
45a-50 Southern Inner Piedmont	**20.90**	**21.87**	**22.08**
45a-90 Southern Inner Piedmont	**60.92**	**64.90**	**66.31**
45b-44 Southern Outer Piedmont	**49.52**	**54.71**	**56.57**
45c-3 Carolina Slate Belt	**27.13**	**33.89**	**38.07**
45d-11 Talladega Upland	**42.32**	**47.63**	**50.64**
45h-1 Pine Mountain Ridges	**68.15**	**71.77**	**74.14**
	Southeastern Plains Ecoregion		
65d-20 Southern Hilly Gulf Coastal Plain	57.23	63.09	63.84
65d-39 Southern Hilly Gulf Coastal Plain	**44.18**	**49.51**	**52.53**
65h-17 Tifton Upland	18.29	22.61	24.11
65k-102 Coastal Plain Red Uplands	36.51	36.96	37.79
65l-184 Atlantic Southern Loam Plains (Vidalia Upland)	26.58	39.20	43.84
65o-23 Tallahassee Hills/ Valdosta Limesink	72.10	75.69	84.08
65o-3 Tallahassee Hills/ Valdosta Limesink	57.46	66.12	70.13
	Blue Ridge Ecoregion		
66d-43 Southern Crystalline Ridges and Mountains	**62.80**	**70.07**	**72.30**
66d-44-2 Southern Crystalline Ridges and Mountains	**68.21**	**74.31**	**79.72**

TABLE 7.1 (CONTINUED)
Mean Metric Scores from Sample Streams after Bootstrap Resampling 25 Times

Stream and Subecoregion	Mean Standardized Metric Score (100 Individuals)	Mean Standardized Metric Score (200 Individuals)	Mean Standardized Metric Score (300 Individuals)
66d-58 Southern Crystalline Ridges and Mountains	**59.41**	**65.47**	69.20
66g-23 Southern Metasedimentary Mountains	**65.78**	**71.36**	72.83
66g-71 Southern Metasedimentary Mountains	64.48	65.92	65.93
66j-19 Broad Basins	**45.78**	**56.77**	**64.68**
66j-23 Broad Basins	**48.27**	**54.66**	**59.01**
66j-25 Broad Basins	**29.77**	**35.97**	**38.88**
66j-26 Broad Basins	**57.41**	**64.49**	**69.65**
66j-28 Broad Basins	**67.20**	**79.33**	**87.84**
Southwestern Appalachians Ecoregion			
68c&d-7 Plateau Escarpment and Southern Table Plateaus	43.72	44.04	44.35
Southern Coastal Plain Ecoregion			
75e-54 Okefenokee Plains	69.90	70.76	70.48
75f-50 Sea Island Flatwoods	59.73	55.33	55.77
75f-95 Sea Island Flatwoods	92.27	91.97	91.72
75h-70 Bacon Terraces	**47.86**	**53.60**	**59.17**

Note: Values in bold showed significant changes with increasing subsample size.

Interquartile variability declined with subsample size and median value increased. Interquartile variability did not overlap among any subsamples.

SUBECOREGION 68C&D–PLATEAU ESCARPMENT AND SOUTHERN TABLE PLATEAUS

68c&d-7

Richness (as Plecoptera taxa) increased slightly with the increase of subsample size. The remaining metric indices did not display any trends over the range of subsample sizes (Figure 7.6 and Table 7.6). Hydropsychidae were absent from all subsamples. Interquartile variability declined substantially with increasing subsample size, however, interquartile variability greatly overlapped among all three subsample sizes (Figure 7.7 and Table 7.7).

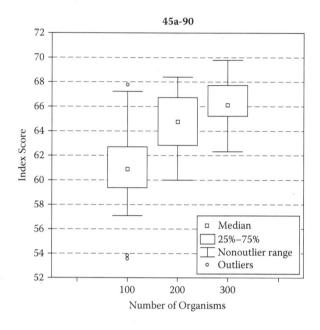

FIGURE 7.2   Macroinvertebrate index score distributions (based upon 25 replicate subsamples) at different subsample sizes in Site 45a-90.

SUBECOREGION 75F–SEA ISLAND FLATWOODS

75f-50

Richness (as Chironomidae taxa), composition (as percent Tanypodinae/total Chironomidae), and tolerance (as tolerant taxa) changed with subsample size. FFG (as percent filterer) index did not change. Oligochaeta and odonata were completely absent from all subsamples. Interquartile variability declined in larger subsamples,

TABLE 7.2
Metric Index Scores before and after Standardization for Site 45a-90

Metric	Raw Score				Standard Score		
	100	200	300	Whole	100	200	300
EPT Taxa	9.08	12.88	14.04	16.00	53.41	75.76	82.59
% Chironomidae Taxa	41.28	39.76	39.73	40.16	54.94	56.98	57.01
% Cricotopus and Chironomus/TC	1.61	1.82	1.66	1.47	95.57	95.01	95.44
NCBI	6.12	5.97	5.96	5.97	61.52	63.45	63.69
% Scraper Taxa	5.76	5.52	5.89	5.71	14.44	13.83	14.77
% Clinger Taxa	54.80	53.98	54.00	54.53	85.63	84.34	84.38
Mean	**19.78**	**19.99**	**20.21**	**20.64**	**60.92**	**64.90**	**66.31**

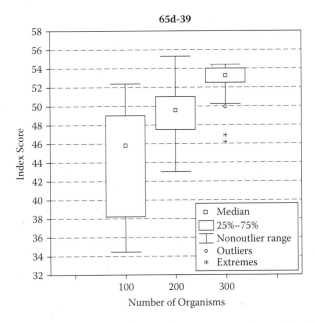

FIGURE 7.3 Macroinvertebrate index score distributions (based upon 25 replicate subsamples) at different subsample sizes in Site 65d-39.

as did median index values. Interquartile variability overlapped among all three subsamples (Figure 7.7 and Table 7.7).

SUBSAMPLING RECOMMENDATION

The mean macroinvertebrate indices of the three levels of subsample sizes for each stream site were paired into three combinations: (1) 100- and 200-organism subsamples; (2) 200- and 275- or 300-organism subsamples; and (3) 100- and 275- or

TABLE 7.3
Metric Index Scores before and after Standardization for Site 65d-39

Metric	Raw Score				Standard Score		
	100	200	300	Whole	100	200	300
Plecoptera Taxa	4.40	6.28	7.04	9.00	79.26	97.19	99.41
Trichoptera Taxa	3.96	4.84	6.32	8.00	61.88	74.88	92.63
% Oligochaeta	0.36	0.26	0.19	0.20	97.14	98.12	98.78
% Hydropsychidae/ Trichoptera	97.36	95.53	96.45	95.83	2.82	4.79	3.81
% Predator	9.88	9.70	9.13	9.27	22.20	21.80	20.52
% Filterer	32.84	31.14	32.73	32.35	1.77	0.29	0.00
Mean	**24.80**	**24.63**	**25.31**	**25.78**	**44.18**	**49.51**	**52.53**

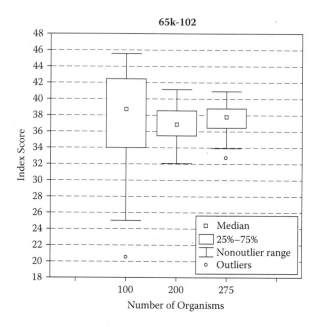

FIGURE 7.4   Macroinvertebrate index score distributions (based upon 25 replicate subsamples) at different subsample sizes in Site 65k-102.

300-organism subsamples. The first combination was used to examine the recommended subsample size of 100 individuals in the original RBP (Plafkin et al. 1989), and also of Georgia Department of Natural Resources (DNR) protocols that have been in use until very recently. This protocol was further examined by testing the third combination of paired subsamples. The second combination of paired subsamples tested the adequacy/inadequacy of 200 individuals as prescribed by the current RBP for stream health assessments (Barbour et al. 1999). Two hundred individuals have been found to be insufficient by previous studies (see, for example, Ostermiller and Hawkins 2004).

TABLE 7.4
Metric Index Scores before and after Standardization for Site 65k-102

Metric	Raw Score				Standard Score		
	100	200	275	Whole	100	200	275
% Tanypodinae/TC	22.82	22.90	22.86	14.37	35.23	35.02	35.13
% Gastropoda	0.00	0.00	0.00	0.00	0.00	0.00	0.00
% Hydropsychidae/ Total Trichoptera	68.49	66.74	67.00	66.67	32.02	33.26	33.00
Scraper Taxa	4.48	4.88	5.00	5.00	90.83	98.00	100.00
% Shredder	3.28	2.78	3.04	2.89	38.05	32.25	35.27
% Collector	21.08	21.31	21.45	39.31	22.90	23.23	23.31
Mean	**20.03**	**19.77**	**19.89**	**21.37**	**36.51**	**36.96**	**37.79**

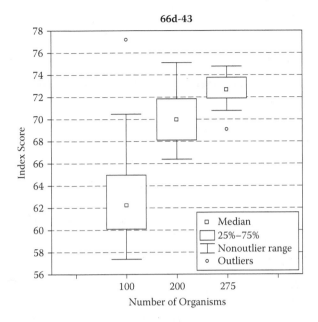

FIGURE 7.5 Macroinvertebrate index score distributions (based upon 25 replicate subsamples) at different subsample sizes in Site 66d-43.

For every pair of subsamples, the least significant range (LSR) value between their mean macroinvertebrate indices was compared to the mean index difference (MID) value. The MID value exceeding the LSR value was considered significantly different (at the 95% confidence level). Significant MID values are bold in Table 7.8, which summarizes the multiple-range tests.

Various metrics have been found to be affected by change in subsample size. For example, in Subecoregion 45–Southern Inner Piedmont, subsample size was found to affect richness (i.e., EPT taxa—number of taxa in the Ephemeroptera, Plecoptera and Trichoptera families) while the remaining metrics were not. Larger subsamples gave

TABLE 7.5
Index Scores before and after Standardization for Site 66d-43

	Raw Score				Standard Score		
Metric	100	200	275	Whole	100	200	275
Diptera Taxa	15.92	23.44	26.52	28.00	53.07	78.13	88.40
% Plecoptera	25.32	24.80	24.76	24.76	82.34	80.65	80.51
% Odonata	0.68	0.60	0.67	0.64	85.09	86.84	85.33
% Dominant Individuals	18.36	18.00	18.02	18.10	27.99	29.49	29.34
% Shredder	27.00	26.30	26.15	26.35	80.48	78.39	75.78
Clinger Taxa	14.44	20.20	22.48	24.00	47.81	66.89	74.44
Mean	**16.95**	**18.89**	**19.77**	**20.31**	**62.80**	**70.07**	**72.30**

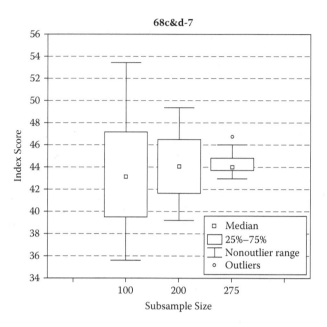

FIGURE 7.6   Macroinvertebrate index score distributions (based upon 25 replicate subsamples) at different subsample sizes in Site 68c&d-7.

higher estimates for richness. Metrics whose indices did not change across the range of subsample sizes were composition, FFG, and habit; all having scored fairly close to the index of the whole sample. Because the number of individuals in each subsample size was closer to the full population size, variability (interquartile range) decreased in larger subsamples. For this site, reliance on information from a 100-organism subsample would lead to erroneous judgment of the stream condition. Using information from 200- or 300-organism subsamples would reduce the chance of making such

TABLE 7.6

Metric Index Scores before and after Standardization for Sites 68c and d-7

Metric	Raw Score				Standard Score		
	100	200	275	Whole	100	200	275
Plecoptera Taxa	1.64	1.84	2.00	2.00	41.00	46.00	50.00
% Hydropsychidae/ Total Trichoptera	0.00	0.00	0.00	0.00	100.00	100.00	100.00
% Tanypodinae/TC	15.12	15.38	15.57	15.53	25.96	23.30	22.24
NCBI	5.16	5.39	5.41	5.40	48.20	49.01	48.46
Scraper Taxa	0.96	1.00	1.00	1.00	14.33	14.93	14.93
% Clinger	11.88	11.20	11.00	11.36	32.81	31.02	30.46
Mean	**5.79**	**5.80**	**5.83**	**5.88**	**43.72**	**44.04**	**44.35**

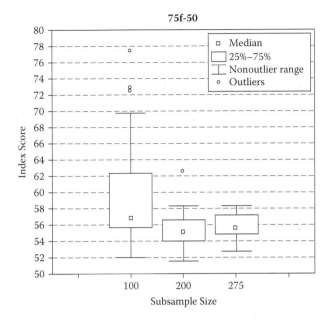

FIGURE 7.7 Macroinvertebrate index score distributions (based upon 25 replicate subsamples) at different subsample sizes in Site 75f-50.

errors. These two subsamples, however, gave similar information on stream condition, as there was no significant difference in their mean indices. Therefore, to save time and unnecessary expense, a subsample with 200 organisms was adequate.

In another stream from the same subecoregion, 45a-50, richness was affected by subsample size. Larger subsamples gave higher estimates of richness. But composition, FFG, tolerance, and habit metrics were not affected by subsample size. A 100-organism subsample gave just as good an estimate as 300-organism subsamples. This suggests that these metrics increased at a proportional rate to increasing subsample size. The observed

TABLE 7.7
Metric Index Scores before and after Standardization for Site 75f-50

	Raw Score				Standard Score		
Metric	100	200	275	Whole	100	200	275
Chironomidae Taxa	2.40	4.60	5.36	6.00	20.25	38.82	45.23
% Oligochaeta	0.00	0.00	0.00	0.00	100.00	100.00	100.00
% Odonata	0.00	0.00	0.00	0.00	100.00	100.00	100.00
% Tanypodinae/TC	55.72	57.54	52.80	53.85	23.43	7.08	6.45
Tolerant Taxa	10.23	13.12	15.00	15.00	30.73	3.62	0.45
% Filterer	3.96	4.34	4.33	4.30	83.99	82.45	82.48
Mean	**12.05**	**13.27**	**12.92**	**13.19**	**59.73**	**55.33**	**55.77**

TABLE 7.8
Multiple-Range Tests of Mean Indices across Subsamples

Site Number	Subsamples N1 and N2		Subsamples N2 and N3		Subsamples N1 and N3	
	LSR	MID	LSR	MID	LSR	MID
45a-35	2.03	**2.37**	2.03	0.12	2.13	**2.25**
45a-50	1.42	0.89	1.42	0.22	1.50	1.11
45a-90	1.43	**3.97**	1.43	**1.54**	1.51	**5.51**
45b-44	2.25	**5.19**	2.25	1.85	2.37	**7.04**
45c-3	2.19	**6.77**	2.19	**4.17**	2.31	**10.94**
45d-11	2.76	**5.31**	2.76	**3.01**	2.90	**8.32**
45h-1	2.43	**3.62**	2.43	2.36	2.56	**5.98**
65d-20	2.05	**5.86**	2.05	0.75	2.16	**6.61**
65d-39	2.23	**5.33**	2.23	**3.01**	2.34	**8.34**
65h-17	1.41	**4.32**	1.41	**1.50**	1.49	**5.82**
65k-102	2.38	0.18	2.38	0.54	2.50	0.72
65l-184	4.93	**12.61**	4.93	3.99	5.19	**16.60**
65o-23	1.87	**3.68**	1.87	**8.18**	1.97	**11.86**
65o-3	2.06	**8.66**	2.06	**4.01**	2.17	**12.67**
66d-43	1.68	**7.27**	1.68	**2.63**	1.77	**9.90**
66d-44-2	2.10	**6.10**	2.10	**5.41**	2.21	**11.51**
66d-58	1.87	**6.26**	1.87	**3.53**	1.97	**9.79**
66g-23	1.8	**5.60**	1.8	1.47	1.90	**7.07**
66g-71	1.59	**4.08**	1.59	**1.91**	1.67	**5.99**
66j-19	2.83	**10.99**	2.83	**7.91**	2.98	**18.90**
66j-23	1.82	**6.38**	1.82	**4.37**	1.91	**10.75**
66j-25	1.85	**6.20**	1.85	**2.91**	1.95	**9.11**
66j-26	2.46	**7.07**	2.46	**5.17**	2.60	**12.24**
66j-28	2.04	**12.14**	2.04	**8.90**	2.15	**21.04**
68c&d-7	1.90	0.16	1.90	0.31	2.00	0.47
75e-54	1.29	0.85	1.22	0.27	1.22	0.58
75f-50	2.48	**4.40**	2.36	0.44	2.36	**3.96**
75f-95	0.55	0.30	0.55	0.25	0.66	0.55
75h-70	2.13	**5.58**	2.13	**5.73**	2.24	**11.31**

Notes: N1 = 100 organisms; N2 = 200 organisms; N3 = 275 or 300 organisms.
MID values in bold are significant at 95% confidence interval.

increase in the sole richness metric was not sufficient to make a significant difference in the overall mean macroinvertebrate index across the range of subsamples.

In an adjacent subecoregion, 45b-44, subsample size affected the estimates for richness, FFG, and habit metrics. One-hundred–organism subsamples gave underestimates for these metrics. Richness was best measured when 200 organisms were used. But for FFG and habit metrics, 275 organisms were required to get better estimates. The seemingly subtle differences in the raw scores of these metrics across subsamples were magnified substantially when equal weights were given to the scores (i.e., standardization).

The overall mean indices of 100- and 200-organism subsamples differed significantly from each other suggesting the latter subsample provides more information (Table 7.8). But the mean index of 200 organisms did not differ significantly from that of 275 organisms and, therefore, the information given by these two subsamples was similar. Hence, a subsample of 200 organisms provided an adequate index of stream condition.

Similar analyses for all subecoregions were conducted by Rai (2006), with various metrics being substantially affected by changes in subsample size and resulting in overall changes in the combined macroinvertebrate indices and interpretation of stream health. The overall recommendations for subsample size, then, are based upon the ability to reduce the variability around the mean value of the composite metric for each subecoregion.

METRIC RESPONSE TO SUBSAMPLE SIZE

In all but one (65h-17) of the 26 study sites that used at least some kind of richness measures, subsample size was found to affect the richness. Biotic richness increased when there were more organisms present in the subsample. This finding is consistent with previous studies (Growns et al. 1997; Cao et al. 1998; Sovell and Vondracek 1999; Duggan et al. 2002). However, Simpson's Diversity Index was the exception. It was not found to be as sensitive to sample size as other metrics of richness, as Simpson's index is weighted toward the abundances of the most common species and responds poorly to the addition of rare species to increase richness (Magurran, 1988; also supported by Veijola et al. 1996). A subsample of 100 organisms was sufficient to estimate Simpson's index. For other richness metrics, the largest subsample (i.e., 275 or 300 organisms in this case) was required. Vinson and Hawkins (1996) also suggested using greater than 300 organisms to obtain more accurate inferences for richness.

In most instances, metrics that utilized percentage community composition or relative abundances did not change when larger numbers of individuals were used in the subsample. Such community metrics were *percent chironomidae, ratio of Cricotopus or Chironomus to total Chironomidae, percent EPT, percent Ephemeroptera, percent Gastropoda, ratio of Hydropsychidae to total Trichoptera, percent Isopoda, percent noninsect macroinvertebrates, percent Odonata, percent Oligochaeta, percent Plecoptera, ratio of Tanypodinae to total Chironomidae, percent Tanytarsini, percent dominant individuals, percent pollution intolerant, percent pollution tolerant, percent clinger, percent collector, percent filterer, percent predator, percent scraper,* and *percent shredder*. Overall, this may be a result of "standardization" of these values on a percentile basis and, then, subsequent standardization when creating the overall macroinvertebrate metric. Increasing organism counts in the subsample resulted in a proportional increase in the respective taxa. A subsample of 100 organisms was equally informative about these metrics as the other two larger subsamples. Similar conclusions for some of the metrics (e.g., *percent EPT abundance, percent dominant taxa*) have been made elsewhere (see, for example, Duggan et al. 2002).

FFG and habit metrics (excluding those describing community percentage and relative abundance) were not consistent across the range subsamples. In general, these values increased with increasing subsample size. FFG metrics describe the dominant feeding mechanisms of biota (Rosenberg and Resh 1996). Metrics such as predator, scraper, and shredder taxa included in this study are sensitive to taxa richness but measure the functioning of the benthic community rather than just the structure. Even though larger subsamples contain a more diverse assemblage (see the section "The Effect of Subsample Size on Metric Scores"), many taxonomically different individuals may exhibit the same feeding pattern and proportionately contribute to the community's dominant trophic character. This may explain why metric scores improved when a greater number of organisms were used in the subsample. The rest of the metrics in the functional feeding group category (percentages of collector, filterer, predator, scraper, and shredder taxa) did not improve with increasing subsample size as previously discussed.

Habit metrics are descriptions of the movement and positioning mechanisms of benthic organisms (Merritt and Cummins 1996). The habit metrics used in this study (clinger, burrower, sprawler, and swimmer taxa) are also sensitive to biotic richness. The higher taxonomic richness in larger subsamples may be responsible for the increase in habit scores, again, because of the possible addition of new taxa having the same habits. Many macroinvertebrates, although taxonomically different, are known to share similar modes of locomotion or to occupy similar types of substrates (Merritt and Cummins 1996). As discussed earlier, the percentage of clinger taxa was the only habit metric to remain unaffected by subsample size.

In the tolerance/intolerance metric category, Hilsenhoff's Biotic Index (HBI) was used in only one study site, whereas the North Carolina Biotic Index (NCBI) was used more often. HBI is a measure of the overall organic-pollution tolerances of taxa present in a community (Hilsenhoff 1987). NCBI is a modified form of HBI and also attempts to measure the tolerance level of biota to other impairments (Lenat 1993). Both indices were found to be insensitive to variation in subsample size. Sovell and Vondracek (1999) came to similar conclusions about HBI, but similar comparisons for the NCBI have not been performed anywhere before. Both biotic indices depend heavily upon richness values. Even though new taxa were added in larger subsamples, as demonstrated by the increased richness in this study, consistencies of HBI and NCBI across subsamples indicate a proportional increase in the ratio of pollution sensitive taxa to insensitive taxa for both indices.

A summary of the minimum required subsample size of both the subecoregion scale and the ecoregion scale is provided in Table 7.9. It is clear that there is no single subsample size that can be relied upon to describe health streams in all subecoregions.

Subecoregions in the Blue Ridge (Site 66) required at least 300 individuals. Subecoregion in the Southwestern Appalachians (Site 68) required only 100 individuals. Subecoregions in the Piedmont (Site 45) required either 200 or 300 individuals. Some subecoregions in the Southeastern Plains (Site 65) and the Coastal Plain (Site 75) required only 100 individuals while others required 200 or even 300 individuals. The general trend seen here indicates that using 300 individuals becomes important for sites in the extreme north Georgia, whereas sites located elsewhere do not always require that many.

At the ecoregion level, 300-organism subsamples were the appropriate sizes to minimize the risk of making erroneous conclusion about stream health. Even though Ecoregion 68 showed that only 100 individuals were necessary, this recommendation

TABLE 7.9

Recommended Minimum Required Sample Size at the Subecoregion and the Ecoregion Scales

Subecoregion	Sample Size	Ecoregion	Sample Size
66	300		
66	300	66	300
66	300		
68c&d	100	68	100
45a	300		
45b	200		
45c	300	45	300
45d	300		
45h	200		
65d	300		
65h	300		
65k	100	65	300
65l	200		
65o	300		
75e	100		
75f	200	75	300
75h	300		

should be treated with some restrictions because only a single subecoregion was studied.

Streams in the Blue Ridge generally had high flow velocity with a high concentration of dissolved oxygen, low ambient water temperature, and diverse habitat types. High macroinvertebrate diversity (richness) is usually associated with such stream conditions; hence, the need for 300 individuals. High gradient streams were also found in some subecoregions of the Piedmont and the Southeastern Plains, but regardless of subecoregion, high-gradient streams generally required a minimum of 300 individuals. Low-gradient streams, on the other hand, did not require that level as often as high-gradient sites.

Another trend was that minimally impaired (reference) sites required at least 300 individuals (Table 7.10). This was because high macroinvertebrate diversity occurs in

TABLE 7.10

Percentage of Total Sites Showing Recommended Subsample Sizes

Site Type	300 Individuals	200 Individuals	100 Individuals	Total
High Gradient	75%	19%	6%	100%
Low Gradient	38%	31%	31%	100%
Reference Condition	75%	12.5%	12.5%	100%
Impaired Condition	52%	29%	19%	100%

streams where there is little or no impairment. But, a surprising 52% of the impaired sites also displayed the need for 300 individuals. However, most of the impaired sites requiring 300 individuals were also high-gradient streams, which may have influenced the results.

The RBP method compares the mean index and interquartile ranges of reference sites to that of test sites whose health conditions are to be determined. To conclude that a test site is impaired, there must be a clear separation between the 25th percentile index value of the reference site and the 75th percentile index value of the test site. An overlap indicates the test site is similar to the reference site and, therefore, in good health. The assumption is that the interquartile range of the reference site represents the true values of healthy streams for the subecoregions. Since our results showed that interquartile range changes across the subsample sizes, an important question arises: Does subsample size affect the ability of reference sites to distinguish themselves from impaired sites? Cao et al. (1998) had expressed concerns about sample sizes of less than 300 individuals cannot effectively use the macroinvertebrate communities' information and may greatly underestimate the differences between reference and impacted sites. The ineffectiveness of small sample size to characterize macroinvertebrate communities were supported by our results. But my examination of Subecoregions 65o and 66g showed that reference sites, regardless of subsample sizes, were still separable from impaired sites (Figure 7.8 and Figure 7.9). However, we did not examine the remaining subecoregions due to the absence of complete data sets (equal sets for both reference and impaired sites); thus, Subecoregions 65o and 66g may be anomalous.

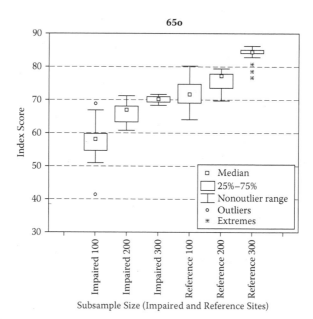

FIGURE 7.8 Macroinvertebrate index score distributions (based upon 25 replicate subsamples) at different subsample sizes between impaired and reference sites for Subecoregion 65o.

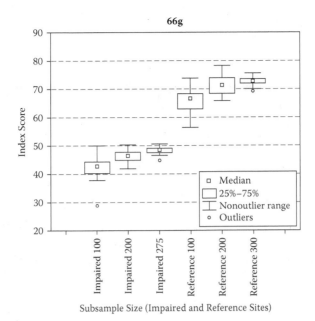

FIGURE 7.9 Macroinvertebrate index score distributions (based upon 25 replicate subsamples) at different subsample sizes between impaired and reference sites for Subecoregion 66g.

Cost effectiveness has always been a central issue in utilizing the RBP, mainly regarding the subsample size, because sorting and taxonomic identification makes up the bulk of the entire process. We found identification to be relatively slow for the initial 100 organisms but progressed rapidly for 200 and 300 organisms because of the recurrence of similar taxa. Cumulative time for processing and identifying 200 organisms from 100 organisms was increased by 78.9%. This is a substantial increase in cost but necessary because 100 organisms were not adequate for 82% of the subecoregions. Cumulative time to identify 300 organisms from 200 organisms was only increased by 39.83%. This increased cost was repaid by the increased ability of metrics to characterize stream health for 59% of the subecoregions, but proved futile for 24% of the subecoregions.

CONCLUSIONS

The performance of the rapid bioassessment metrics recommended for the ecoregions and subecoregions of Georgia and examined in this study was variable in terms of sensitivity to subsample size. Richness metrics were most sensitive and increased with increasing subsample size. At times, the increase was large enough to substantially affect the overall mean macroinvertebrate index value even without the compounding effects of other sensitive metrics. To a lesser extent, some FFG metrics (scraper, predator, and shredder taxa) and habit metrics (burrower, clinger, sprawler, and swimmer taxa) increased in value with increasing subsample size, and so did HBI and NCBI. Other metrics did not show any consistent trends.

The study has led us to conclude that the previously recommended subsample sizes of 100 organisms and 200 organisms were not adequate to characterize stream conditions for all subecoregions. Three-hundred–organism subsample sizes were necessary for subecoregions of northern Georgia, whereas for those in middle and southern Georgia that level was not always required. Stream gradient was also an important factor because high-gradient sites mostly required 300 individuals, whereas most low gradient sites did not. Every subecoregion, with its distinct geographical conditions, influences the streams and the macroinvertebrate community in its own way and this was reflected in the difficulty of determining one common subsample size to fit all subecoregions. Certainly, subsampling 300 organisms would circumvent this problem (as it did at the ecoregion level), but that would mean unnecessarily increasing spending for the evaluation of some subecoregions. Therefore, we recommend using individual subecoregional subsample sizes for specific subecoregions because they provide adequate characterization of stream conditions and a more cost-effective approach. There was some evidence that subsample size does not affect the ability of reference sites to differentiate from impaired sites. This provides further support that any of the three recommended subsample sizes appropriate to their respective subecoregion can detect the difference between reference and impaired sites.

We recommend the following changes and additions:

1. Equal number of sites should be analyzed for each subecoregion for a robust comparison. We used samples collected for the Georgia Ecoregions Project and, due to sampling problems beyond the control of this study, were unable to acquire equal number of sites for each ecoregion.
2. Subsample sizes with more than 300 individuals should be investigated. Studies using only a few metrics have shown that even 300 organisms may not be enough to characterize stream conditions. We recommend using all metrics and not just richness because FFG and tolerance metrics are sensitive to subsample size as well.
3. Questions regarding the subsample size effects on reference sites to differentiate from impaired sites should also be explored for all subecoregions. This can be done by analyzing an equal number of impaired and reference sites for each subecoregion. We were able to examine this question for only two subecoregions.
4. Stream gradients and associated benthic communities seemed to be correlated with effectiveness of subsample size. Further studies could examine affects of stream gradients on subsample size and the ability to differentiate reference and highly impaired sites.

Finally, we would like to emphasize the usefulness of the RBP in the monitoring, management, and restoration of streams. At its best, RBP provides important biological, physical, and chemical characteristics of a stream in a quick manner. Based on this information, agencies and interested parties will be able to identify and prioritize issues in their decision-making process and, at its worst, the RBP may provide inaccurate information and consequently mislead management efforts. Our conclusions on subsample size requirements will help minimize costly mistakes for resource managers and decision makers.

REFERENCES

Barbour, M.T., J. Gerritsen, B.D. Snyder, and J.B. Stribling. 1999. *Rapid Bioassessment Protocols for Use in Streams and Wadeable Rivers: Periphyton, Benthic Macroinvertebrates and Fish,* 2nd ed. EPA 841-B-99-002. Washington, DC: U.S. Environmental Protection Agency, Office of Water.

Cao, Y., D.D. Williams, and D.P. Larsen. 2002. Comparison of ecological communities: The problem of sample representativeness. *Ecological Monographs* 72(1): 41–56.

Cao, Y., D.D. Williams, and N.E. Williams. 1998. How important are rare species in aquatic community ecology and bioassessment? *Limnology and Oceanography* 43(7): 140–1409.

Carter, J.L., and V.H. Resh. 2001. After site selection and before data analysis: Sampling, sorting, and laboratory procedures used in stream benthic macroinvertebrate monitoring programs by USA state agencies. *Journal of the North American Benthological Society* 20:658–682.

Caton, L.W. 1991. Improved subsampling methods for the EPA "Rapid Bioassessment" benthic protocols. *Bulletin of the North American Benthological Society* 8(3): 317–319.

Courtemanch, D.L. 1996. Commentary on the subsampling procedures used for rapid bioassessment. *Journal of the North American Benthological Society* 15(3): 381–385.

Duggan, I.C., K.J. Collier, and P.W. Lambert. 2002. Evaluation of invertebrate biometrics and the influence of subsample size using data from some Westland, New Zealand, lowland streams. *New Zealand Journal of Marine and Freshwater Research* 36: 117–128.

Efron, B., and R.J. Tibshirani. 1993. *An Introduction to the Bootstrap.* New York: Chapman & Hall.

Gore, J.A., J.R. Olson, D.L. Hughes, and P.M Brossett. 2005. *Reference Conditions for Wadeable Streams in Georgia with a Multimetric Index for the Bioassessment and Discrimination of Reference and Impaired Streams* (Ecoregion Reference Site Project, Phase II, Final Report). Atlanta: Georgia Department of Natural Resources.

Growns, J.E., B.C. Chessman, P.K. McEvoy, and I.A. Wright. 1995. Rapid assessment of rivers using macroinvertebrates: Case studies in the Napean River and Blue Mountains, NSW. *Australian Journal of Ecology* 20: 130–141.

Growns, J.E., B.C. Chessman, J.E. Jackson, and D.G. Ross. 1997. Rapid assessment of Australian rivers using macroinvertebrates: Cost and efficiency of 6 methods of sample processing. *Journal of the North American Benthological Society* 16(3): 682–693.

Hilsenhoff, W.L. 1987. An improved biotic index of organic stream pollution. *Great Lakes Entomologist* 20(1): 31–40.

Lenat, D.R. 1993. A biotic index for the southeastern United States: Derivation and list of tolerance values, with criteria for assigning water-quality ratings. *Journal of North American Benthological Society* 12(3): 279–290.

MacArthur, R.H., and E.O. Wilson. 1967. *The Theory of Island Biogeography.* Princeton, NJ: Princeton University Press.

Magurran, A.E. 1988. *Ecological Diversity and Its Measurement.* London: Croom Helm.

Merritt, R.W., and K.W. Cummins (eds.). 1996. *An Introduction to the Aquatic Insects of North America,* 3rd ed. Dubuque, IA: Kendall/Hunt.

Metzelling, L., and J. Miller. 2001. Evaluation of the sample size used for the rapid bioassessment of rivers using macroinvertebrates. *Hydrobiologia* 444: 159–170.

Ostermiller, J.D., and C.P. Hawkins. 2004. Effects of sampling error on bioassessments of stream ecosystems: Application to RIVPACS-type models. *Journal of North American Benthological Society* 23(2): 363–382.

Plafkin, J.L., M.T. Barbour, K.D. Porter, S.K. Gross, and R.M. Hughes. 1989. *Rapid Bioassessment Protocols for Use in Streams and Rivers: Benthic Macroinvertebrates and Fish.* EPA 440-4-89-001. Washington, DC: U.S. Environmental Protection Agency, Office of Water Regulations and Standards.

Rai, U. 2006. The effect of sample size on rapid bioassessment scores and management efficiency. MS Thesis. Columbus State University, Columbus, GA.

Rosenberg, D.M., and V.H. Resh 1996. Use of aquatic insects in biomonitoring. In *An Introduction to the Aquatic Insects of North America,* 3rd ed., eds. R.W. Merritt and K.W. Cummins, 87–98. Dubuque, IA: Kendall/Hunt.

Somers, K.M., R.A. Reid, and S.M. David. 1998. Rapid biological assessments: How many animals are enough? *Journal of the North American Benthological Society* 17(3): 348–358.

Sovell, L.A., and B. Vondracek. 1999. Evaluation of the fixed-count method for Rapid Bioassessment Protocol III with benthic macroinvertebrate metrics. *Journal of the North American Benthological Society* 18(3): 420–426.

Steel, R.G.D., and J.T. Torrie, 1960. *Principles and Procedures of Statistics.* New York: McGraw-Hill.

Veijola, H., J.M. Jarmo, and V. Marttila. 1996. Sample size in the monitoring of benthic macrofauna in the profundal of lakes: Evaluation of the precision of estimates. *Hydrobiologia* 322: 301–315.

Vinson, M.R., and C.P. Hawkins, 1996. Effects of sampling area and subsampling procedure on comparisons of taxa richness among streams. *Journal of the North American Benthological Society* 15(3): 392–399.

8 Taxonomic Resolution and Cost Effectiveness of Rapid Bioassessment

Jodi A. Williams, James A. Gore, and Michele P. Brossett

CONTENTS

Qualitative approaches using benthic macroinvertebrate assemblages, such as rapid assessment approaches, have recently been accepted as a means to identify water quality problems due to point and nonpoint source pollution, and to document long-term regional changes in water quality (Barbour et al. 1999). Rapid assessments reduce effort and associated costs in evaluating a site in relation to quantitative techniques by (1) reducing the number of habitats sampled and replicating sample units taken per habitat; (2) collecting less silt and particulate matter making sorting faster and easier; (3) considering only a fraction of the animals collected, thus reducing time spent on identification; and (4) identifying organisms to family or higher taxonomic level. Rapid assessment approaches can also provide summary information of study sites in a manner that can be understood by nonspecialists such as managers, the general public, and decision makers (Resh and Jackson 1993). This form of water quality analysis is accomplished by expressing analytical measures (metrics) as single scores and then placing the scores in categories of varying water quality based on regional background data. The EPA's Rapid Bioassessment Protocol (RBP), although not strictly qualitative, is one method that is frequently employed because elements of both qualitative and quantitative approaches are working in conjunction so that results are achieved in a timely manner.

Although rapid bioassessments are efficient, not all bioassessment protocols allow for identification of organisms to family or higher taxonomic levels. There is considerable debate about the taxonomic resolution necessary to accurately determine community condition in bioassessments. Resh and Unzicker (1975) have demonstrated that component species of 61 of the 89 genera for which water-quality tolerances have been established fall into different tolerance categories. They stress the importance of species level or "lowest practical level" due to the substantial variation among species within genera and families and their different responses when exposed to various kinds of pollution. Hawkins et al. (2000) concluded that, in taxonomically rich areas, it was necessary to identify to genus or species level to explain variation among communities, but in areas of little taxonomic diversity, family-level identification was sufficient.

Although a diverse benthic fauna in streams suggest the need for generic or specific levels of identification, it has not been determined whether the aquatic ecosystems in taxonomically rich regions respond to stressors more consistently at those taxonomic levels. Greater variation in species from site to site may reduce the ability to detect a deviation from the unimpaired or minimally impaired stream (i.e., the reference condition), and information gained from the genus level may represent ecological noise, depending upon the specificity of the benthic community response to stress (Bailey et al. 2001). In addition, species-level identification is not always possible because immature stages are collected and species designations are based upon the morphological characteristics of adult insects or larval-adult associations (Lenat and Resh 2001).

With varying taxonomic resolution, Bowman and Bailey (1997) found little effect on multivariate descriptions of variation among communities, particularly when comparing reference to impaired sites. They argued that sufficient resolution for sensitive and accurate bioassessments is achieved when organisms are identified to family level or higher.

To be effective in evaluation of stream impairment, organisms identified to genus or species level must provide significantly more descriptive information than family level, and they must enable better detection of departure from reference condition or the resources expended on taxonomic identification will not be cost effective (Bailey et al. 2001).

As the RBP continues to increase in application across the United States, it will become necessary to resolve the issue of taxonomic resolution, not only to assess the sensitivity of the assessment, but also to make recommendations to state agencies regarding the costs and benefits of recommended identification levels. For example, during the process of making recommendations to states, the representatives from a state may determine that economic cost outweighs resolution in determining "stream health" or impairment. That is, states may be willing to accept a greater chance of error in exchange for greater geographical coverage.

In this chapter, we present the results of our research to determine if departure of macroinvertebrate metric scores from the reference condition was easier to detect with generic or specific identifications than identification of the same sample of individuals to the family level alone. As in other analyses presented in this text, we used data from the Georgia Ecoregions Project (Gore et al. 2005) for this analysis. Since all individuals were identified to the lowest possible taxonomic level, most commonly genus or species, recombining these data to the familial level allowed us to easily examine taxonomic resolution.

Samples, each containing approximately 200 individuals (as required by the RBP), were collected from 31 sites. To represent a spectrum of the diversity in underlying geology and geography of the state of Georgia, which ultimately affects benthic communities, 10 samples from the Blue Ridge Mountain ecoregion and the Piedmont ecoregion, as well as 11 samples from the Southeastern Plains ecoregion, were examined. Because drought conditions affected the number of streams that could be sampled in some regions of the state, we chose only those regions from which we could draw at least 10 sampled sites. Thus, we selected the Southern Metasedimentary Mountains, the

Southern Inner Piedmont, and the Sand Hills subecoregions to complete the analysis. Within each group, at least 5 samples represented reference conditions and remaining samples were considered to have some level of impairment.

Metrics incorporated into the subecoregional indices were used to evaluate taxonomic resolution at the lowest practical level (LPL), at the generic level, and at the familial level. When compiling invertebrate indices, the metrics that did not apply at the generic and familial identification levels were considered invalid and were omitted from the index. Index discrimination efficiencies were computed by using the 25th percentile of reference condition scores. A 25th percentile is considered sufficiently conservative to protect aquatic resources and still allow for some uncertainty of reference condition sites (Jessup and Gerritsen 2000). The 25th percentile of reference condition scores is used as a threshold value for management action since impairment measurements, as an index score, fall along a continuum. A threshold reflects the risk and uncertainty of misclassification of stream health: the risk of declaring a good stream as impaired (Type I error) and the risk of declaring an impaired stream as good (Type II error) (Jessup and Gerritsen 2000). Box-and-whisker plots were used to exhibit distribution of reference condition and impaired index scores and for evaluating taxonomic resolution requirements.

To perform cost/benefit analyses, a stopwatch was used to time taxonomic resolution performance for 12 samples. Times recorded were standardized and averaged so that total time spent on identification at each taxonomic level and total time spent on mounting chironomid larvae within each subecoregion represented "costs." Costs versus "benefits," degree of information reflected by high discrimination efficiencies, were compared at each taxonomic level. The level of taxonomic performance exhibiting the greatest discrimination efficiency between reference and impaired sites with the subecoregions was determined to be the most economical means for accurately classifying stream water quality.

Thirty-one benthic macroinvertebrate samples from five reference sites and five or six impaired sites from three subecoregions in Georgia were collected during the index period (September through February of 2000, 2001, and 2002). Of the 6782 macroinvertebrates identified, 427 were identified to family level; 3613 to genus level; 2557 to species level; and 185 to subfamily, class, or tribe.

Metrics included within the subecoregion-specific invertebrate indices exhibited greater discrimination efficiency (DE) at different levels of taxonomic resolution (Table 8.1). In some cases, discrimination efficiencies were greater at the familial level than at the species or genus level and in some cases the results were quite mixed. However, if there was variability within a metric, the greatest discrimination efficiency was obtained at the LPL level.

Overall index scores and discrimination efficiencies varied at each taxonomic level within each subecoregion. Discrimination efficiencies were 100% for both the lowest practical level and generic-level indices within the Southern Inner Piedmont subecoregion; the family level index had a discrimination efficiency of 80% (Table 8.1). The Sand Hills subecoregion had the greatest variation among the three subecoregions with discrimination efficiencies at the family level, generic level, and LPL of 50%, 83%, and 67%, respectively (Table 8.2); whereas the discrimination efficiencies for the Southern Metasedimenatary Mountains subecoregion were 100% for the three indices (Table 8.3).

TABLE 8.1
Metric Categories, Metrics Complied into Invertebrate Indices (Hughes 2006), and Metric Discrimination Efficiencies (DEs) for Three Levels of Taxonomic Resolution for Three Subecoregions

Southern Inner Piedmont

Metric Category	Metric	LPL DE	Genus DE	Family DE
Richness	EPT Taxa	60%	60%	80%
Composition	Percent Chironomidae	100%	100%	100%
Composition	*Cricotopus* and *Chironomus*/ TC	100%	100%	*
Tolerance	Percent Tolerant	100%	100%	40%
Trophic/FFG	Percent Scraper	80%	60%	60%
Habit	Burrower Taxa	40%	0%	40%

Sand Hills

Richness	Trichoptera Taxa	50%	50%	50%
Composition	Percent Trichoptera	50%	50%	50%
Composition	*Cricotopus* and *Chironomus*/ TC	50%	50%	*
Tolerance	Tolerant Taxa	67%	67%	67%
Trophic/FFG	Percent Scraper	83%	0%	67%
Habit	Clinger Taxa	33%	33%	33%

Southern Metasedimentary Mountains

Richness	EPT Taxa	80%	80%	100%
Composition	Percent Chironomidae	80%	80%	80%
Composition	Percent Tanypodinae/ TC	80%	*	*
Tolerance	Percent Dominant	80%	60%	100%
Tolerance	NCBI	60%	60%	*
Trophic/FFG	Scraper Taxa	100%	20%	100%
Habit	Burrower Taxa	60%	100%	60%

Note: DE values marked with an asterisk (*) are considered to be invalid for the purposes of analysis. TC = Total Chironomidae.

TABLE 8.2
Stream Index Scores and Index Discrimination Efficiencies (DEs) for the Southern Inner Piedmont

Station ID	LPL Index	Genus Index	Family Index
	Impaired Sites		
45a-35	71	66	84
45a-50	23	41	27
45a-59	30	51	33
45a-61	22	38	42
45a-90	59	65	59
	Reference Sites		
45a-03	77	84	80
45a-3	66	79	63
45a-89	79	63	70
HH16	77	67	75
HH18	72	84	78
Composite DE	**100%**	**100%**	**100%**

TABLE 8.3
Stream Index Scores and Index Discrimination Efficiencies (DEs) for the Sand Hills

Station ID	LPL Index	Genus Index	Family Index
	Impaired Sites		
65c-12	58	65	47
65c-3	71	70	64
65c-4	17	15	26
65c-40	69	81	67
65c-8	55	71	66
65c-88	34	41	32
	Reference Sites		
65c-80	69	76	59
65c-89	67	67	52
HH24	78	96	79
HH25	91	90	92
HH26	59	72	55
Composite DE	**67%**	**83%**	**50%**

TABLE 8.4
Stream Index Scores and Index Discrimination Efficiencies (DEs) for the Southern Metasedimentary Mountains

Station ID	LPL Index	Genus Index	Family Index
	Impaired Sites		
66g-30	25	29	40
66g-31	35	45	55
66g-42	59	58	46
66g-44	37	27	24
66g-71	45	55	58
	Reference Sites		
66g-2	63	57	56
66g-2-2	70	61	75
66g-23	83	87	87
66g-5	83	74	81
66g-6	74	80	80
Composite DE	100%	100%	100%

Identifications to LPL in the Southern Inner Piedmont, to generic level in the Sand Hills, and to familial level in the Southern Metasedimentary Mountains (Table 8.4) displayed the greatest discriminatory efficiency for classifying reference and impaired sites.

TAXONOMIC EFFORT AND COST

For 12 impaired sites, approximately 41 hours were spent mounting chironomid larvae while identification of all macroinvertebrates, including the chironomid larvae, to LPL required approximately 32 hours and 45 minutes. Time spent on each level of identification, for all taxonomic orders, averaged 30 minutes for familial identification, 1.75 hours for generic level identification, and 30 minutes for species identification. Average time spent on identification is the cumulative total for each taxonomic level, in this case 2.75 hours. By combining cumulative average identification time with average mounting time, cumulative time spent in processing the samples averaged 6.5 hours (Table 8.5).

Processing time varies depending upon the taxonomic composition of the samples and those taxa that dominate various stream ecosystems. For example, in the southern part of Georgia, streams tend to be dominated by a high diversity of chironomidae larvae, which require additional preparation time, while streams in northern Georgia tend to be dominated by larger (and more easily identifiable) mayflies, stoneflies, and caddisflies. Thus, when identification times and mounting times were combined, average total time spent on taxonomy was approximately 4.5 hours per sample for the Southern Metasedimentary Mountains subecoregion, 5.25 hours per sample for the Southern Inner Piedmont subecoregion, and 8.75 hours per sample for the Sand Hills subecoregion.

TABLE 8.5

Sample Preparation Time (Slide Mounting) and Taxonomic Identification Time for 12 Samples

Taxonomic Level	ID Time (hrs)	Mount Time (hrs)	Average ID Time (hrs)	Average Mount Time (hrs)	Average Total ID Time (hrs)
Family	~4.50	0	~0.50	0	~0.50
Genus	~21.00	~41.0	~1.75	~3.50	~5.25
Species	~7.50	0	~0.50	0	~0.50
Total (LPL)	~32.75	~41.0	~2.75	~3.50	~6.25

TAXONOMIC RESOLUTION ANALYSIS

Subecoregion-specific indices for the Southern Inner Piedmont, the Sand Hills, and the Southern Metasedimentary Mountains were evaluated at the familial level, generic level, and LPL for taxonomic resolution and their effectiveness on predicting stream health. Based upon the discriminatory ability of the indices and depictions of score distributions, analysis indicated that taxonomic requirements varied among subecoregions (Figure 8.1 through Figure 8.6). We predict that this general trend would occur in any ecoregional analysis at any geographic location.

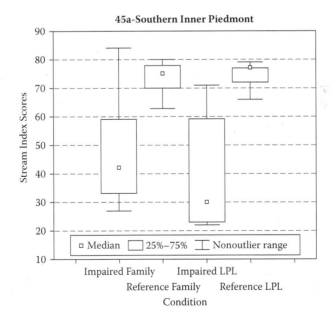

FIGURE 8.1 Southern Inner Piedmont box-and-whisker plots exhibiting the distribution of reference condition and impaired stream index scores for the family index as compared to the LPL (often a mix of taxonomic levels) score distributions.

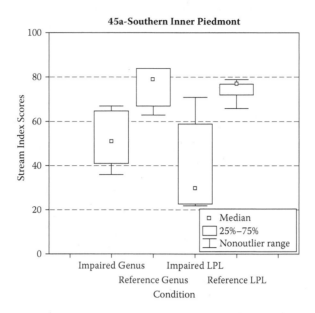

FIGURE 8.2   Southern Inner Piedmont box-and-whisker plots exhibiting the distribution of reference condition and impaired stream index scores for the genus index as compared to the LPL (often a mix of taxonomic levels) score distributions.

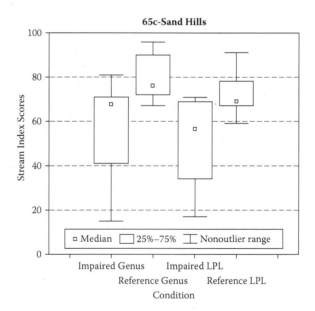

FIGURE 8.3   Sand Hills box-and-whisker plots exhibiting the distribution of reference condition and impaired stream index scores for the genus indices as compared to the LPL (often a mix of taxonomic levels) score distributions.

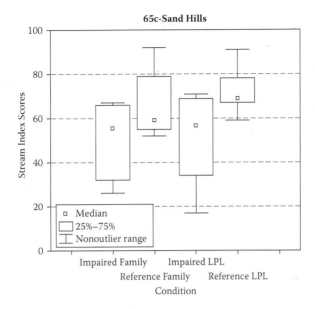

FIGURE 8.4 Sand Hills box-and-whisker plots exhibiting the distribution of reference condition and impaired stream index scores for the family level indices as compared to the LPL (often a mix of taxonomic levels) score distributions.

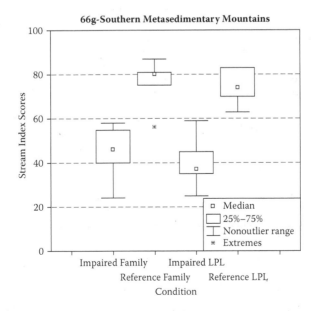

FIGURE 8.5 Southern Metasedimentary Mountains box-and-whisker plots exhibiting the distribution of reference condition and impaired stream index scores for the family indices as compared to the LPL (often a mix of taxonomic levels) score distributions.

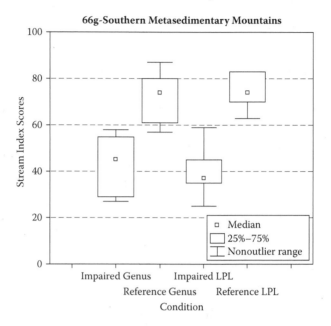

FIGURE 8.6   Southern Metasedimentary Mountains box-and-whisker plots exhibiting the distribution of reference condition and impaired stream index scores for the generic indices as compared to the LPL (often a mix of taxonomic levels) score distributions.

The invertebrate index for the Southern Metasedimentary Mountains had a discrimination efficiency of 100% when metric scores for the three levels of taxonomic identification were computed. LPL index scores exhibited slightly greater discrimination potential than familial index scores due to the slightly greater degree of separation between interquartile ranges of reference condition and impaired stream index scores. The variance among reference condition scores was less than reference condition scores in the LPL index and 100% of reference condition index scores were above the 25th percentile threshold. One outlier (which may indicate natural variability, misclassification of stream conditions *a priori*, or an underrepresented site class; Jessup and Gerritsen 2000) was found in the family index, but removing that score from the index did not affect the discriminatory efficiency. Because the degrees of overlap between familial and LPL discriminatory efficiencies were small, identification to either LPL or familial level is adequate for this subecoregion.

By examining the amount of overlap in box-and-whisker plots, similar recommendations for taxonomic resolution requirements can be made for any ecoregion. For example, identification to LPL was recommended for the Southern Inner Piedmont subecoregion since discrimination efficiencies were 80% at the family level, slightly better at the generic level, and 100% at LPL. In the Sand Hills ecoregion, discrimination efficiencies were the greatest using generic-level identifications.

Taxonomic resolution requirements may also be reflective of the physical and biological conditions within each ecoregion, and the composition and diversity of the benthic communities in those streams. For example, the Southern Metasedimentary Mountains is an area of open low hills, with some rugged, isolated mountains such as the Cohutta Mountains and Fort Mountain. Anthropogenic activity is somewhat limited within this region; thus, stress imposed on aquatic ecosystems may not be as great as in other more flat or urban regions. In these hillier regions, Ephemeroptera, Plecoptera, and Trichoptera (EPT) taxa are known to exist and dominate some communities of unimpaired streams. EPT responses to stressors have been well documented and tolerances well established, allowing a greater ease in separation of stressed conditions, creating greater discrimination efficiency at family and generic levels. Relative to those streams dominated by a high diversity of Chironomid larvae, these EPT-dominated streams also had lower overall diversity; these findings support similar studies (Hawkins et al. 2000) that indicate that family-level identifications are sufficient. In contrast, biotic diversity was highest in the Southern Inner Piedmont, which might explain identification requirements to LPL. As Resh and Unzicker (1975) reported, established water quality tolerance values for 61 of 89 genera fall into different tolerance categories, indicating that identification to LPL is necessary to explain the variation among species and their response to various stressors.

COST/BENEFIT ANALYSIS

As stipulated in the RBP, all organisms were identified to LPL. This level of identification entails a fair amount of investment in time and in money, but identification to LPL is needed before taxonomic resolution analysis can be performed, and it is possible that this level of identification is the only level that adequately discriminates between reference condition and impaired streams. Once taxonomic resolution requirements have been predetermined, cost/benefit analyses can be used to examine not only the real costs of taxonomic work but also potential savings that would be realized when future benthic work is needed for assessment and monitoring programs.

When identification time below the family level is required, the time needed to mount various larval representatives must also be included in the commitment to overall analysis. This will significantly increase the investment per sample but will be necessary to improve identification and separation of reference and impaired streams, using the appropriate level metrics. Table 8.6 summarizes the "costs" (using a base salary of $10 per hour) of taxonomic effort in each of the subecoregions.

Taxonomic resolution requirements vary among subecoregions. Our limited analysis suggests that taxonomic requirements will vary within all ecoregions and subecoregions. This type of analysis should be completed before additional assessment or monitoring takes place if costs are to be minimized, as the risks of misclassifying stream health can be high. If classification results in Type I errors and healthy streams (similar to reference conditions) are classified as impaired, additional time and money will be spent on unnecessary assessments. If classification results in Type II errors and impaired streams are classified as healthy, costs become even greater because stream health deterioration is compounded over time, which means

TABLE 8.6
Cost/Benefit Analysis (per Sample), Based upon a Rate of $10 per Hour for Taxonomic Analysis

	Southern Inner Piedmont (45a)	Sand Hills (65c)	Southern Metasedimentary Mountains (66g)
Family Level Costs	$ 5.00	$ 7.50	$ 5.00
Genus Level Costs	$12.50	$27.50	$12.50
Species Level Costs	$ 2.50	$ 5.00	$ 5.00
Combined Level Identification Costs	$20.00	$40.00	$22.50
Mounting Costs (Chironomidae)	$35.00	$47.50	$22.50
Total Costs per Sample	**$55.00**	**$87.50**	**$45.00**
Less Costs per Sample (Family)			**$ 5.00**
Less Costs per Sample (Genus)		**$82.50**	
Future Net Savings per Sample		**$ 5.00**	**$40.00**

more assessments and frequent monitoring in the future (not to mention increased chances of total maximum daily load [TMDL] assessments or fines).

Therefore, by predetermining taxonomic resolution requirements and by performing cost/benefit analyses, agencies involved in assessment and monitoring programs would not only minimize risks of misclassifying stream health, they could also identify regions that may require less taxonomic effort and economic benefits, which can be transferred to rehabilitation and restoration of impaired stream ecosystems.

REFERENCES

Bailey, R.C., R.H. Norris, and T.B. Reynoldson. 2001. Taxonomic resolution to benthic macroinvertebrate communities in bioassessments. *Journal of the North American Benthological Society* 20: 280–286.

Barbour, M.T., J. Gerritsen, B.D. Snyder, and J.B. Stribling. 1999. *Rapid Bioassessment Protocols for Use in Streams and Wadeable Rivers: Periphyton, Benthic Macroinvertebrates and Fish*, 2nd ed. EPA 841-B-99-002. Washington, DC: U.S. Environmental Protection Agency, Office of Water.

Bowman, M.F., and R.C. Bailey. 1997. Does taxonomic resolution affect the multivariate description of the structure of freshwater benthic macroinvertebrate communities? *Canadian Journal of Fisheries and Aquatic Sciences* 54: 1802–1807.

Gore, J.A., J.R. Olson, D.L. Hughes, and P.M. Borssett. 2005. *Reference Condition for Wadeable Streams in Georgia with a Multimetric Index for the Bioassessment and Discrimination of Reference and Impaired Streams* (Ecoregions Reference Sites Project, Phase II Final Report). U.S. Environmental Protection Agency, Clean Water Act, Section 319(h) FY 98–Element 1. Atlanta: Georgia Department of Natural Resources.

Hawkins, C.P., R.H. Norris, J.N. Hogue, and J.W. Feminella. 2000. Development and evaluation of predictive models for measuring the biological integrity of streams. *Ecological Applications* 10: 1456–1477.

Hughes, D.L. 2006. Development of biological reference conditions of wadeable streams in the major ecoregions and subecoregions of Georgia. MS Thesis. Columbus State University, Columbus, GA.

Jessup, B.K., and J. Gerritsen. 2000. *Development of a Multimetric Index for Biological Assessment of Idaho Streams Using Benthic Macroinvertebrates*. Prepared for the Idaho Department of Environmental Quality. Owings Mills, MD: Tetra Tech, Inc.

Lenat, D.R., and V.H. Resh. 2001. Taxonomy and stream ecology: The benefits of genus- and species-level identifications. *Journal of the North American Benthological Society* 20: 287–298.

Resh, V.H., and J.K. Jackson. 1993. Rapid assessment approaches to biomonitoring using benthic macroinvertebrates. In *Freshwater Biomonitoring and Benthic Macroinvertebrates*, eds. D.M. Rosenberg and V.H. Resh, 195–233. London: Chapman & Hall.

Resh, V.H., and J.D. Unzicker. 1975. Water quality monitoring and aquatic organisms: The importance of species identification. *Journal of the Water Pollution Control Federation* 47: 9–19.

9 Quality Assurance/ Quality Control

What Does It Reveal about the Reliability of the Rapid Bioassessment Protocol?

Tracy J. Ferring, James A. Gore,
and Duncan L. Hughes

CONTENTS

Traditionally, biomonitoring had been utilized to quantify "before-and-after" impacts from a known disturbance. Biomonitoring, as currently implemented, can be used to predict impacts to an aquatic system prior to major anthropogenic impairments within a watershed (Rosenberg and Snow 1977), as well as to serve as a template, ensuring compliance to statutory requirements as set by the Environmental Protection Agency (EPA) through the Clean Water Act (CWA). For compliance measures, biological criteria can be applied to evaluate the effects of effluent discharges or other human-induced changes within a catchment and to document that water quality standards have, or have not, been violated (Roper 1985).

The theory behind the use of the ecoregion concept is that adjoining land forms with similar geologic features, soil types, vegetation, and climatic influences will most likely possess similar biological communities (Omernik and Gallant 1990; Hughes 1995; Omernik 1995). This concept is useful in conjunction with bioassessment programs because it can be used to characterize and predict natural variations among systems within similar geographic regions, as well as to detect responses to disturbances based on some reference condition (Hughes and Larsen 1988).

With the variable geology and vegetation patterns across any large geographic area, it should be expected that a variety of macroinvertebrate assemblages will reflect the ambient water quality and habitat structure of those systems. Similarly, any degradation of habitat and deviation from typical water quality in a region should be reflected by changes in the composition of the macroinvertebrate community. Characterizing a representative macroinvertebrate community in minimally impaired catchments serves as a reference point for other stream ecosystems that have been subjected to some sort of anthropogenic stress.

A reference condition, as prescribed by bioassessment protocols, is defined as "the condition that is representative of a group of minimally disturbed sites organized by physical, chemical and biological characteristics" (Reynoldson et al. 1997). The biological condition of a stream, or group of streams, that are classified as "reference" then serve as the point of comparison for all other streams within a catchment or ecoregion. The chemical, physical, and biological attributes of a reference stream can then be used to identify levels of impairment in streams that are known to be altered. The differences between the biological condition of a reference and impaired site can be quantified through a series of biological metrics. These metrics are then used to develop a ranking system to identify streams that have acceptable or degraded water quality per EPA standards. To accurately assess the effects of anthropogenic influences, natural variability within these geographical boundaries must be characterized. As an example, in Georgia there is a very distinctive geological, vegetative, and geomorphological transition from the northwest region to the southeast region. This change in ecoregional character dictates a variety of stream morphologies with variable habitat structures and water chemistries. The final determination of a series of metrics must somehow account for natural biological variability within and across ecoregional boundaries (Mississippi Department of Environmental Quality [MDEQ] 2003).

We have decided to examine the "representativeness" of stream samples, when collected according to the EPA protocol, using quality assurance/quality control (QA/QC) samples for comparison. In conjunction with sampling proposed reference and impaired sites, additional samples were collected as dictated by the Quality Assurance Project Plan (QAPP) (Columbus State University [CSU] 2000). Throughout each phase of the Georgia Ecoregion Study, there were a number of duplicate samples taken to satisfy the QA/QC requirements of the QAPP. These duplicate samples were taken to assess the repeatability and precision of the collected data, as well as to assess the training and level of effort between and among field teams. We assessed QC data in terms of measurement quality objectives (MQOs) as outlined by the QAPP, but addressed, more specifically, the amount and degree of variability present in the wadeable stream ecosystems across the state of Georgia, and how these samples affect initial characterization of the reference condition.

There were two designations for QC samples collected to satisfy the QAPP document: (1) a spatial, "duplicate reach" QC sample, and (2) a temporal, "phase" QC sample. According to QAPP procedures, the QC type of duplicate sampling is performed to assess the precision and accuracy of the field teams and the representativeness of the data as some measure of "data quality" in bioassessment programs. It is important to analyze the consistency of field teams to ensure that personnel are properly trained so that the collection of biological data are free from bias and error, but more important, the objective of analyzing the additional biological data was to determine if the sites chosen to characterize the biological condition were true representations of the biological community in that stream.

During characterization of the reference condition for Georgia ecoregions and subecoregions, the additional data collected through the QC samples were not used in the creation of overall metric scores or in characterization of the final biological index. Thus, it became important to determine how well the randomly assigned samples from the selected stream reach accurately reflected the composition of the macroinvertebrate community. There have been numerous studies of the effect of sample size on the variability of biotic indices in bioassessment programs (Li et al. 2001; Metzeling and Miller 2001; as well as information presented earlier). Increase in sample size will result in the increase of number of individuals collected, but, more important, also correspond to an increase in the number of taxa in the system being sampled. It has also been demonstrated that increasing the size of the sampling area (whether it be sampling more than one riffle or a combination of habitats to constitute one sample) has an effect on the range of variance (Norris et al. 1993; Hannaford and Resh 1995). When considering bioassessment protocols, it raises the question of determining what important taxa may have been be missed and how these excluded taxa may influence the range of variability of the metrics used to determine the reference and impaired condition.

There have been a number of papers analyzing variability in data using Rapid Bioassessment Protocols (RBPs; see reviews by Hannaford and Resh 1995), but the majority of these have centered on specific habitat types such as riffles and runs (see, for example, Feminella 2000) and are based upon the assumption that swifter water habitats yield the highest species richness and abundance of invertebrates (Hynes 1970; Allan 1995). Also common in these previous studies has been the use of "in-field" subsampling of macroinvertebrates as the basis for characterizing variability in the data sets (Metzeling and Miller 2001). Logically there is some question about bias resulting from *in-field* subsampling of macroinvertebrates, as there may be a tendency to choose the larger, more obvious organisms for analysis, resulting in skewed final metrics and biotic indices calculated for a stream. Additionally, given that many macroinvertebrates have very specific habitat requirements, it is to be expected that metric results would vary as a function of the range of particular habitats being sampled. There are numerous species that thrive in habitats such as tree roots along stream banks and woody debris (i.e., snags), an especially productive habitat type in low-gradient stream systems typical of the southern United States, for example (Benke et al. 1985). The sampling of multiple habitats in bioassessment protocols provides a better biological *picture* of the faunal communities that are subject to changes in habitat structure and water quality.

We attempted to address a number of questions. With regard to the Georgia ecoregions QC data, does the inclusion of additional taxonomic data change the range of variability and the definition of the reference or impaired condition? Second, will the restriction or expansion of those ranges of variability create difficulties in interpretation of anthropogenic stressors on the biotic community? Likewise, does the range of variability within the identified metrics and biotic indices hinder the decision-making process for water resource managers? The answers to these questions might indicate that increasing the sample size (for example, increasing the number of reference and impaired site samples, and increasing the reach length) in RPB programs may better characterize natural variability within and between ecoregions, and also reduce the variability of the final metrics used to characterize the reference condition and water quality, as well as more narrowly defining numerical criteria of stream health (Gore et al. 2005).

The QAPP describes the procedures that were used in data collection and their rationale, as well as a series of activities and reporting procedures that were used to document data quality. As prescribed by the QAPP document for the Georgia Ecoregions Project, a number of sites were designated for additional sampling. To address QC/QA protocols related to data quality, 10% of all the designated sampling sites were required to have duplicate sampling performed. These duplicate samples fell into two designations: "spatial" (200 meter QC) and "temporal" (phase QC).

Quality control samples that were designated as spatial essentially "doubled" the length of the reach designated for sampling. Once the primary sample reach of 100 meters was established, and all RBP sampling requirements satisfied, the immediate upstream 100 meter reach was sampled.

Temporal QCs were sites that were sampled in succeeding phases of the ecoregions project, where the originally established sample site reach was resampled in a subsequent "index period." This sampling approach addressed two possible variations within a stream ecosystem: (1) the variability of the distribution of habitats longitudinally within a catchment and (2) changes of the macroinvertebrate communities over time.

All QC sites were randomly chosen; as a result, there was some unevenness in the number of duplicate reference- and impaired-site samples, as well as the number of spatial and temporal QC samples collected. Additionally, the total number of QC sites collected for this project was not evenly distributed throughout each ecoregion and subecoregion.

As sites were sampled (original and QA/QC sites) and taxonomic identifications were completed, all of the physical, chemical, and biological data gathered were entered into the Ecological Data Application System (EDAS) database for further analysis. As previously documented (Gore et al. 2005), metric values calculated by EDAS were separated by ecoregion and subecoregion, as well as by impairment status, within the ecoregional designation. These raw metric scores were initially used to distinguish which metrics were to be considered as candidates for the final biotic index. A series of analyses was performed to assess the ability of each metric to discriminate between the characteristics of a "reference" stream and an "impaired" stream. For those sites designated as reference streams, interquartile ranges of the distribution of metric values for each metric category were calculated. These percentile values then served as templates for the calculations of discrimination efficiency, metric score standardization, and the final biotic index for each ecoregion and subecoregion.

There are many important factors to consider when developing biocriteria for stream ecosystems. In the process of collecting biological data, field methods cannot predict if the information being collected is an accurate portrayal of the ecosystem under investigation (Intergovernmental Task Force on Monitoring Water Quality [ITFM] 1995). The properties of a given field sample may be known, but typically biological data are collected with the intent of answering questions relating to much larger spatial and temporal scales (Barbour et al. 1999). The consistency of field methods and level of effort in collecting biological data is the key to obtaining information that is representative of field conditions, at that point in time, but truly accurate assessments of the biological data are hindered because the natural variability of the ecosystem cannot be controlled (Resh and Jackson 1993).

In a similar manner, as the stepwise process for identifying an ecoregional biotic index from raw metric values, the QC samples collected for this project were considered for their precision and representativeness of the biological condition. With the RBPs use of multimetric assessment methods, the precision of the total bioassessment score is as important as the precision of the individual metrics that comprise the score (Diamond et al. 1996). Typically, when considering wide-scale bioassessment programs, some form of criterion is established to assess the quality of the data that has been collected. These criteria are commonly referred to as data quality objectives (DQOs) and measurement quality objectives (MQOs). These are qualitative and quantitative parameters developed by those who will use the data to evaluate the bioassessment score's accuracy and how the quality of the data will affect management decisions.

Analysis centered on three measurement parameters associated with the collection of benthic macroinvertebrates: metric values, standardized metric scores, and bioassessment scores. Not only were these parameters analyzed with regard to the prescribed MQOs, but the QC data were also evaluated for their relevance and value to characterizing the reference and impaired conditions, as defined by the biotic index.

Typical evaluations of data repeatability and data quality center on the use of a series of calculations that quantify variability between measures. The following relationships are typical of bioassessment protocols that define measures of acceptable variability (EPA 1995; Barbour et al. 1999; Stribling and Bressler 2004):

- Relative percent difference (RPD): Quantifies the proportional difference between two measures as

$$RPD = [(C_1 - C_2)/(C_1 + C_2)] \times 100$$

  where C_1 is the larger of the two values being compared, and C_2 is the smaller of the two values being compared (Berger et al. 1996).
- Root mean square of error (RMSE): Used as an estimate of the standard deviation of a group of observations. The RMSE is determined by performing an analysis of variance between duplicate samples to determine the mean square error (MSE) that is representative of within-group variance.
- Coefficient of variability (CV): Calculated by expressing the standard deviation as a percentage of the mean. The coefficient of variability for a population was calculated as

$$CV = (RMSE/Y) \times 100$$

where Y was the mean of the dependent variable (e.g., metric values, scores, etc.).

TABLE 9.1

Precision Measurement Quality Objectives for Benthic Macroinvertebrates as Defined in the Georgia Ecoregions Project Quality Assurance Project Plan (QAPP)

Measurement Parameter	Precision Level
Metric Values	RPD < 20%
	RMSE = TBD
Metric Scores	RPD < 5%
	RMSE = TBD
Bioassessment Scores	RPD < 5%
	RMSE = TBD

Source: See also Barbour et al. (1999); USEPA (1995).
Note: TBD = to be determined. RMSE levels developed as a result of this study.

Values associated with RPDs and RMSEs characterized some level of precision among the parameters being analyzed. As defined by the QAPP for the Georgia Ecoregions Project, RPDs for metric values, metric scores, and bioassessment scores were defined to indicate some level of data quality (see Table 9.1). Additionally, RMSEs for these same measurements parameters were developed here. For each raw metric value and metric score, both RPDs and RMSEs were calculated, while only the candidate metrics used for the development of the biotic index were examined. These calculations provided some indication of not only the quality of the data collected, but also acted as a measure of how representative the biological data were to each ecoregion. This evaluation of the QC data for this project provides a framework for data users and water resource managers to assess the reliability and inherent variability of the proposed biotic indices for the state of Georgia.

Precision calculations of the measurement parameters for benthic macroinvertebrate metrics are presented at the primary ecoregional level. For each ecoregion, calculations of RPDs and RMSE (as required by the QAPP document), and CVs are provided as averages for all QC samples collected (both reference/impaired and spatial/temporal), inclusive of their ecoregional designation. Average RPDs, RMSEs, and CVs are provided for each raw metric value, standardized metric score, and the final bioassessment scores (biotic indices) developed for each ecoregion. Additionally, average RPD, RMSE, and CV calculations are presented relative to the specific metrics used in the developed biotic indices for each primary ecoregion and subecoregion (Gore et al. 2005).

At the subecoregional level, average RPD, RMSE, and CV values for raw metric values, standardized metric scores, and bioassessment scores are provided in the associated appendices. Also included in Appendix G for each ecoregion are RPD, RMSE, and CV calculations (at the ecoregion and subecoregion level), for each stream class (i.e., reference versus impaired), and collectively for all original and QC sites sampled within the ecoregional designation. Therefore, the averages specific to reference or impaired streams in each ecoregional designation only include data relevant to the designated stream class, whereas the averages provided for the entire ecoregion are inclusive of all original and associated QC samples in the ecoregion designation, both reference and impaired, and spatial and temporal.

In conjunction with the required precision measures dictated by the QAPP, RPD values were calculated to compare stream classes, (i.e., reference versus impaired), as well as QC sample type (i.e., spatial versus temporal). Again, these parameters are summarized for all raw metric values at the ecoregional level.

RELATIVE PERCENT DIFFERENCE (RPD) PRECISION MEASURES FOR RAW METRIC VALUES AND STANDARDIZED METRIC SCORES

The raw metric values were analyzed for RPDs, acting both as a measure of data precision and data uncertainty due to natural variability of the lotic ecosystem. Table 9.2 contains a summary of RPDs averaged for all metrics within each category. The average RPDs for individual metrics in each metric category, in most cases, were higher than the measurement quality objectives dictated by the ecoregions QAPP. The RPDs of raw metric values from duplicate reaches were expected to be in 80% agreement (Table 9.2). This is better illustrated in Figure 9.1, which demonstrates that RPD values are relatively consistent between the metric categories and ecoregional designation.

RPDs were also calculated for the standardized metric scores and have been summarized for the primary ecoregions of Georgia in Table 9.3 and illustrated in Figure 9.2. Again, these values are higher than the prescribed MQO of 95% agreement for standardized metric scores.

TABLE 9.2
Average Relative Percent Difference (RPD) for All Raw Metric Values within Each Metric Category and per Primary Ecoregion Designation

Metric Group	Ecoregion					
	45	65	66	67	68	75
Taxonomic Richness	20.9	21.4	20.3	19.1	19.2	18.1
Community Composition	34.2	32.7	32.5	41.5	27.9	30.5
Tolerance/Intolerance	18.0	21.3	19.0	18.0	8.0	18.5
Functional Feeding Group	25.5	24.4	18.5	26.1	13.9	37.2
Life Habit	27.7	29.2	25.7	23.8	26.9	28.4

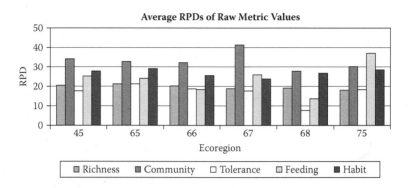

FIGURE 9.1   Average relative percent difference (RPD) for all raw metrics values within each metric category and per ecoregion designation.

TABLE 9.3
Average Relative Percent Difference (RPD) for All Standardized Metric Scores within Each Metric Category and per Primary Ecoregion Designation

	Ecoregion					
Metric Group	**45**	**65**	**66**	**67**	**68**	**75**
Taxonomic Richness	18.7	19.1	17.8	16.2	17.3	17.3
Community Composition	28.2	23.8	28.0	25.7	25.2	20.5
Tolerance/Intolerance	23.1	18.0	26.1	11.6	18.7	20.7
Functional Feeding Group	23.4	22.7	18.6	28.1	8.2	34.2
Life Habit	27.4	28.2	24.0	24.0	25.3	29.4

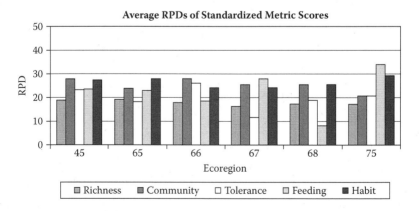

FIGURE 9.2   Average relative percent difference (RPD) for all standardized metric scores within each metric category and per primary ecoregion designation.

TABLE 9.4
Average Root Mean Square Error (RMSE) for All Raw Metric Values within Each Metric Category and per Primary Ecoregion Designation

	Ecoregion					
Metric Group	45	65	66	67	68	75
Taxonomic Richness	3.5	4.0	4.0	3.6	1.1	2.2
Community Composition	13.5	14.3	9.5	11.8	3.7	14.8
Tolerance/Intolerance	10.9	11.7	7.4	9.1	3.5	11.7
Functional Feeding Group	8.7	7.6	5.3	7.5	2.0	6.0
Life Habit	6.0	4.8	4.3	5.1	1.4	2.2

ROOT MEAN SQUARE OF ERROR (RMSE) PRECISION MEASURES OF RAW METRIC VALUES AND STANDARDIZED METRIC SCORES

Another precision measurement utilized was the RMSE, which is a representation of within-group variance, and acts as an estimate of the standard deviation of each population of metric values. Acceptable levels of error associated with RMSEs have not been established or quantified. The values presented here establish the ranges of variability on an ecoregional and subecoregional basis. Similar to the precision measurements for the RPDs of raw metric values and standardized metric scores, the values for RMSEs presented are averages of all metrics within each metric category for each primary ecoregion of Georgia. Subecoregional averages of RMSE values are presented in the associated appendices. The average RMSE was calculated for all raw metrics values as summarized on Table 9.4 and illustrated in Figure 9.3. Similarly, average RMSEs for standardized metric scores are presented in Table 9.5 and illustrated in Figure 9.4.

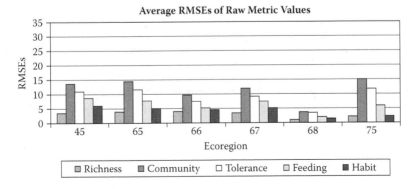

FIGURE 9.3 Average root mean square error (RMSE) for all raw metric values within each metric category and per primary ecoregion designation.

TABLE 9.5

Average Root Mean Square of Error (RMSE) for All Standardized Metric Scores within Each Metric Category per Primary Ecoregion Designation

Metric Group	Ecoregion					
	45	65	66	67	68	75
Taxonomic Richness	21.3	26.1	20.1	22.6	10.2	27.7
Community Composition	25.3	26.8	23.7	23.8	8.8	26.1
Tolerance/Intolerance	26.0	25.4	26.7	16.0	7.5	29.9
Functional Feeding Group	26.0	27.0	21.4	24.2	6.2	31.7
Life Habit	25.3	28.6	22.9	25.6	20.1	31.7

COEFFICIENT OF VARIABILITY (CV) PRECISION MEASURES OF RAW METRIC VALUES AND STANDARDIZED METRIC SCORES

The CV is another measure of variability and precision calculated for raw metric values, standardized metric scores, and bioassessment scores. CV values were calculated to further illustrate the ranges of variability of metrics within and between each ecoregion. Statistically, as the CV value increases, the precision of the variable examined declines. CV values were calculated for raw metric values (presented in Table 9.6 and Figure 9.5) and for standardized metric values (presented in Table 9.7 and Figure 9.6).

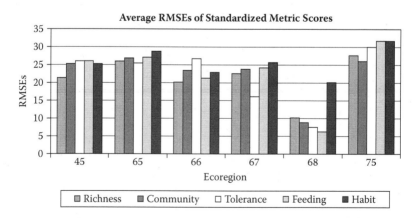

FIGURE 9.4   Average root mean square error (RMSE) for all standardized metric scores within each metric category per primary ecoregion designation.

TABLE 9.6
Average Coefficient of Variability (CV) for All Raw Metric Values within Each Metric Category per Primary Ecoregion Designation

Metric Group	Ecoregion					
	45	65	66	67	68	75
Taxonomic Richness	57.6	69.4	38.2	47.1	27.1	73.1
Community Composition	157.1	156.9	90.3	120.3	39.5	153.3
Tolerance/Intolerance	77.0	61.8	39.3	43.1	11.3	74.7
Functional Feeding Group	68.5	71.3	39.0	54.4	19.6	97.3
Life Habit	60.8	69.5	47.1	38.1	38.1	103.9

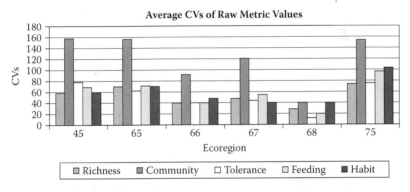

FIGURE 9.5   Average coefficient of variability (CV) for all raw metric values within each metric category and per primary ecoregion designation.

TABLE 9.7
Average Coefficient of Variability (CV) for All Standardized Metric Values within Each Metric Category and per Primary Ecoregion Designation

Metric Group	Ecoregion					
	45	65	66	67	68	75
Taxonomic Richness	49.6	60.7	34.8	39.2	22.9	68.7
Community Composition	94.0	89.2	60.9	74.1	35.7	77.8
Tolerance/Intolerance	56.2	49.6	44.2	24.2	25.6	71.6
Functional Feeding Group	57.1	61.4	37.8	48.6	11.6	78.8
Life Habit	58.5	65.9	43.3	45.3	35.8	96.7

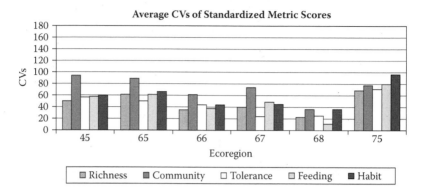

FIGURE 9.6 Average coefficient of variability (CV) for all standardized metric scores within each metric category and per primary ecoregion designation.

PRECISION MEASURES FOR BIOASSESSMENT SCORES

The final precision measures for evaluation of the variability of metrics within and between ecoregional designations centers on the final bioassessment scores that constitute the biotic indices developed for each ecoregion. The average values calculated for RPDs, RMSEs, and CVs presented in Table 9.8 were inclusive of only those metrics that were determined to be indicative of community assemblages that exhibited responses to anthropogenic stress and were descriptive of the reference and impaired condition. In conjunction with Table 9.8, Figure 9.7 illustrates the ecoregional averages of RPD, RMSE, and CV values for the final bioassessment scores used in the development of the biotic index. Comparisons of RPD, RMSE, and CV values for final bioassessment metrics for each ecoregion and their corresponding subecoregions are also presented in Appendix G and summarized in Figure G.1 through Figure G.54.

TABLE 9.8
Average Relative Percent Difference (RPD), Root Mean Square Error (RMSE), and Coefficient of Variability (CV) Values for Final Bioassessment Scores Used in the Development of Biotic Indices for the Primary Ecoregions of Georgia

	Ecoregion					
	45	**65**	**66**	**67**	**68**	**75**
RPDs	20.3	10.1	12.7	10.4	6.7	5.4
RMSEs	18.9	17.4	15.4	14.1	6.0	9.6
CVs	56.0	40.0	23.2	24.3	9.4	13.3

Note: Averages are inclusive of only the metrics used to develop the final biotic index for each ecoregion designation.

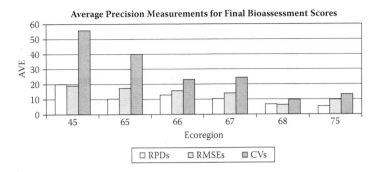

FIGURE 9.7   Average relative percent difference (RPD), root mean square of error (RMSE), and coefficient of variability (CV) values for final bioassessment scores used in the development of biotic indices for the primary ecoregions of Georgia. (Averages are inclusive of only the metrics used to develop the final biotic index for each ecoregion designation.)

COMPARISON OF PRECISION MEASURES AND DISCRIMINATION EFFICIENCIES OF ECOREGIONAL AND SUBECOREGIONAL BIOTIC INDICES

The final biotic indices, developed from the Georgia Ecoregions Project at the ecoregional and subecoregional level, are presented in the following tables. These biotic indices were taken from the Georgia Ecoregions Project numerical index report (Gore et al. 2005). In addition to Tables 9.9 and 9.10 that summarize the metrics used in the biotic indices for the primary ecoregions of Georgia, corresponding averages of the precision measures of RPDs, RMSEs, and CVs are also presented for the standardized metric scores that comprised the final additive bioassessment scores. Along with the precision measures, discrimination efficiency values specific to each metric are described in Appendix G.

Figure 9.8 illustrates the comparison of the precision measures of RPD, RMSE, and CV for the biotic indices of Ecoregion 45 and its subecoregions. Figure 9.9

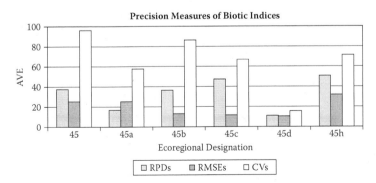

FIGURE 9.8   Average relative percent difference (RPD), root mean square of error (RMSE), and coefficient of variability (CV) values for final bioassessment scores used in the development of biotic indices for Ecoregion 45 and its subecoregions. (Averages are inclusive of only the metrics used to develop the final biotic index for each ecoregion designation.)

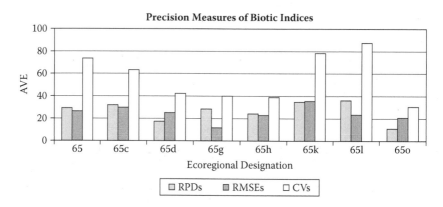

FIGURE 9.9   Average relative percent difference (RPD), root mean square of error (RMSE), and coefficient of variability (CV) values for final bioassessment scores used in the development of biotic indices for Ecoregion 65 and its subecoregions. (Averages are inclusive of only the metrics used to develop the final biotic index for each ecoregion designation.)

illustrates the comparison of the precision measures of RPD, RMSE, and CV for the biotic indices of Ecoregion 65 and its subecoregions. Figure 9.10 illustrates the comparison of the precision measures of RPD, RMSE, and CV for the biotic indices of Ecoregion 66 and its subecoregions. Figure 9.11 illustrates the comparison of the precision measures of RPD, RMSE, and CV for the biotic indices of Ecoregion 67 and its subecoregions, as well as Ecoregion 68 which consisted solely of one Subecoregion—68c&d. Figure 9.12 illustrates the comparison of the precision measures of RPD, RMSE, and CV for the biotic indices of Ecoregion 75 and its subecoregions.

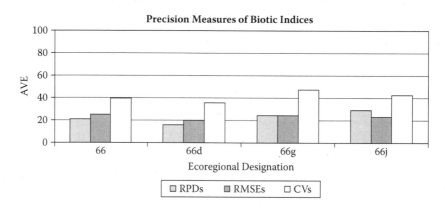

FIGURE 9.10   Average relative percent difference (RPD), root mean square of error (RMSE), and coefficient of variability (CV) values for final bioassessment scores used in the development of biotic indices for Ecoregion 66 and its subecoregions. (Averages are inclusive of only the metrics used to develop the final biotic index for each ecoregion designation.)

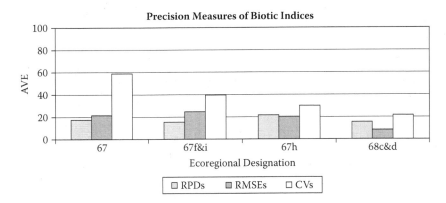

FIGURE 9.11 Average relative percent difference (RPD), root mean square of error (RMSE), and coefficient of variability (CV) values for final bioassessment scores used in the development of biotic indices for Ecoregion 67 and its subecoregions, and Ecoregion 68 which consisted of one subecoregion (c&d). (Averages are inclusive of only the metrics used to develop the final biotic index for each ecoregion designation.)

COMPARISON OF RELATIVE PERCENT DIFFERENCE BY STREAM CLASS DESIGNATION

Although the final determination of metrics that represented the biological condition was ultimately based on each metric's ability to distinguish differences in the characteristics of a reference or impaired stream ecosystem, it was interesting to note the measures of variability of the stream classes themselves. Specifically, the precision measure of relative percent difference was considered to illustrate the variability of raw metric values calculated for reference and impaired streams separately.

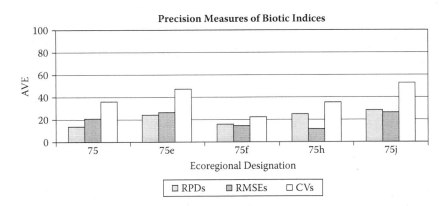

FIGURE 9.12 Average relative percent difference (RPD), root mean square of error (RMSE), and coefficient of variability (CV) values for final bioassessment scores used in the development of biotic indices for Ecoregion 75 and its subecoregions. (Averages are inclusive of only the metrics used to develop the final biotic index for each ecoregion designation.)

TABLE 9.9

Average Relative Percent Difference (RPD) of Quality Control (QC) Sites per Stream Class and per Primary Ecoregion

Metric Group	Class	Ecoregion					
		45	65	66	67	68	75
Taxonomic	Reference	19.0	22.7	21.3	19.7	na	8.6
Richness	Impaired	22.6	19.5	18.9	13.0	19.2	21.9
Community	Reference	35.5	31.6	31.6	42.5	na	26.7
Composition	Impaired	32.0	34.3	33.6	32.5	27.9	32.0
Tolerance/	Reference	16.8	21.0	20.5	18.6	na	20.3
Intolerance	Impaired	19.1	21.8	17.0	12.3	8.0	17.8
Functional	Reference	22.4	26.4	21.4	27.7	na	26.5
Feeding Group	Impaired	28.1	21.4	14.6	11.9	13.9	41.4
Life Habit	Reference	23.8	26.5	29.8	24.7	na	8.6
	Impaired	31.2	33.1	20.3	15.7	26.9	36.3

Note: Values are averaged for all raw metric values within the metric group category. na = No QC sample collected for the stream class designation.

The RPD precision measures of raw metric values for reference and impaired stream classes are presented at the ecoregional level and per metric group are summarized in Table 9.9. Additionally, the differences in variability between the stream classes are illustrated in two manners: (1) each metric category is compared individually between ecoregional designations and (2) each metric category is compared to one another per primary ecoregion designation. The RPD values presented in Table 9.9 are illustrated in Figure 9.13 through Figure 9.23, indicating variability for each metric category in relation to other metric categories, as well as the variability of each metric category within each ecoregional designation.

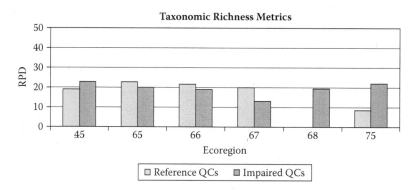

FIGURE 9.13   Comparison of relative percent difference (RPD) values averaged for all raw metric values of the taxonomic richness metrics per ecoregion designation and for stream class quality control (QC) samples. (No reference QC samples were collected for Ecoregion 68.)

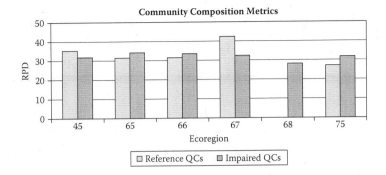

FIGURE 9.14   Comparison of relative percent difference (RPD) values averaged for all raw metric values of the community composition metrics per ecoregion designation and for stream class quality control (QC) samples. (No reference QC samples were collected for Ecoregion 68.)

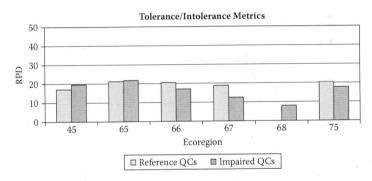

FIGURE 9.15   Comparison of relative percent difference (RPD) values averaged for all raw metric values of the tolerant/intolerant individuals metrics per ecoregion designation and for stream class quality control (QC) samples. (No reference QC samples were collected for Ecoregion 68.)

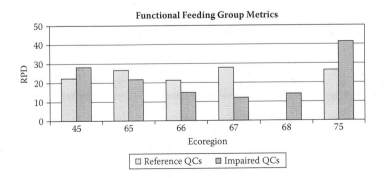

FIGURE 9.16   Comparison of relative percent difference (RPD) values averaged for all raw metric values of the functional feeding group metrics per ecoregion designation and for stream class quality control (QC) samples. (No reference QC samples were collected for Ecoregion 68.)

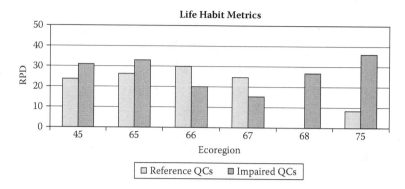

FIGURE 9.17 Comparison of relative percent difference (RPD) values averaged for all raw metric values of the life habit metrics per ecoregion designation and for stream class quality control (QC) samples. (No reference QC samples were collected for Ecoregion 68.)

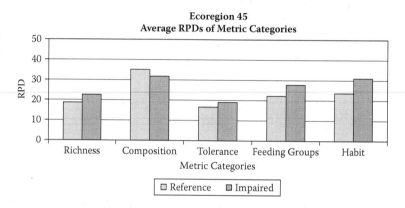

FIGURE 9.18 Comparison of relative percent difference (RPD) values averaged for all raw metric values by stream class designation per metric category for primary Ecoregion 45.

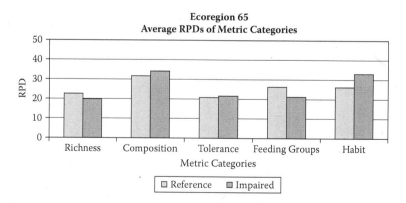

FIGURE 9.19 Comparison of relative percent difference (RPD) values averaged for all raw metric values by stream class designation per metric category for primary Ecoregion 65.

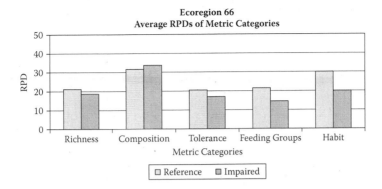

FIGURE 9.20 Comparison of relative percent difference (RPD) values averaged for all raw metric values by stream class designation per metric category for primary Ecoregion 66.

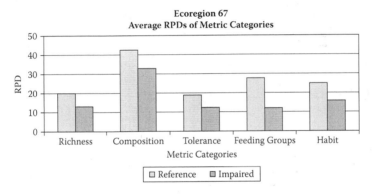

FIGURE 9.21 Comparison of relative percent difference (RPD) values averaged for all raw metric values by stream class designation per metric category for primary Ecoregion 67.

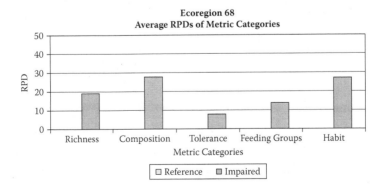

FIGURE 9.22 Comparison of relative percent difference (RPD) values averaged for all raw metric values by stream class designation per metric category for primary Ecoregion 68. (No reference QC samples were collected for Ecoregion 68.)

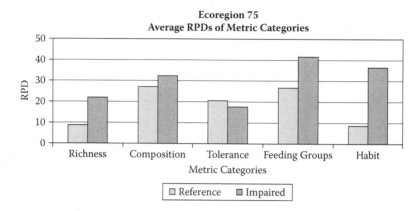

FIGURE 9.23 Comparison of relative percent difference (RPD) values averaged for all raw metric values by stream class designation per metric category for primary Ecoregion 75.

COMPARISON OF RELATIVE PERCENT DIFFERENCE FOR SPATIAL AND TEMPORAL QC SAMPLES

Similar to the analysis of average RPD of raw metric values for all metrics within each metric category and per stream class designation, differences in RPDs between spatial and temporal QC samples were also considered. RPD values are summarized in Table 9.10 and are also illustrated in corresponding Figure 9.24 through Figure 9.34. Again, these values are illustrated in two manners: (1) each metric category is compared

TABLE 9.10

Average Relative Percent Difference (RPD) of Quality Control (QC) Sites per QC Type and per Primary Ecoregion

| Metric Group | QC Type | Ecoregion | | | | | |
		45	65	66	67	68	75
Taxonomic	Spatial	20.1	17.7	10.3	16.9	19.2	13.0
Richness	Temporal	27.4	27.9	21.1	20.6	na	23.5
Community	Spatial	32.1	30.5	22.9	30.0	27.9	19.8
Composition	Temporal	44.0	36.7	33.0	47.1	na	41.5
Tolerance/	Spatial	15.9	21.2	7.4	11.9	8.0	13.5
Intolerance	Temporal	29.9	21.6	20.6	19.4	na	21.3
Functional	Spatial	22.6	22.9	15.8	13.0	13.9	35.3
Feeding Group	Temporal	33.7	26.9	19.5	31.4	na	40.4
Life Habit	Spatial	32.1	36.4	27.2	19.5	32.1	20.2
	Temporal	32.7	30.8	35.1	34.5	na	39.7

Note: Values are averaged for all raw metric values within the metric group category. na = No QC sample collected for the QC type designation.

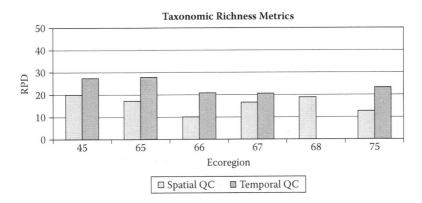

FIGURE 9.24 Comparison of relative percent difference (RPD) values averaged for all raw metric values of the taxonomic richness metrics per ecoregion designation and by quality control (QC) sample designation. (No temporal QC samples were collected for Ecoregion 68.)

individually between ecoregional designations, and (2) each metric category is compared to one another per primary ecoregion designation, illustrating the variability for each metric category in relation to other metric categories, as well as the variability of each metric category within each ecoregional designation.

Rapid bioassessment is essentially a "biological shortcut" in impact assessment studies, where the goal is to sample a wide range of aquatic biota with the fastest methodology (Metzeling and Miller 2001). The underlying premise of rapid bioassessment is the *minimal* effort needed to characterize macroinvertebrate communities that result in *maximum* information. At this point in the evolution of bioassessment protocols, minimal effort is reflected by the limited number of sample replicates and a limitation on the number of collected organisms to be used in metric calculations

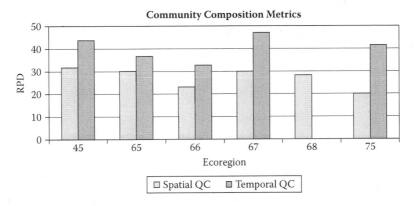

FIGURE 9.25 Comparison of relative percent difference (RPD) values averaged for all raw metric values of the community composition metrics per ecoregion designation and by quality control (QC) sample designation. (No temporal QC samples were collected for Ecoregion 68.)

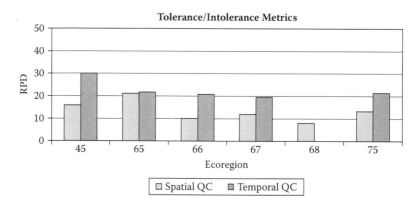

FIGURE 9.26 Comparison of relative percent difference (RPD) values averaged for all raw metric values of the tolerant/intolerant individuals metrics per ecoregion designation and by quality control (QC) sample designation. (No temporal QC samples were collected for Ecoregion 68.)

(Metzeling and Miller 2001). Although stream conditions determined by biological assessments are relayed to the general public and water resource mangers as narrative descriptions (i.e., reference versus impaired lotic systems, or rankings of *good*, *fair*, or *poor*), the final determination of the biological condition is the result of quantitative, numerical indicators with decision thresholds.

Measurement errors in an ecoregional study, due to its complexity and high level of effort, can be compounded from outdated land-use data, as well as errors in field sampling, laboratory subsampling, taxonomic identification and enumeration, data entry, and final metric calculations. The accumulation of errors from these multiple sources results in uncertainty and overall variability (Clark and Whitfield 1994; Diamond et al. 1996). Calculations of variance within the biological parameters

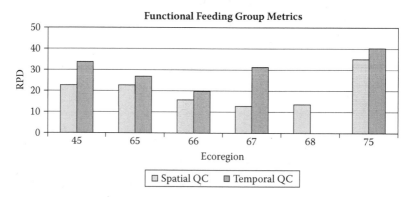

FIGURE 9.27 Comparison of relative percent difference (RPD) values averaged for all raw metric values of the functional feeding group metrics per ecoregion designation and by quality control (QC) sample designation. (No temporal QC samples were collected for Ecoregion 68.)

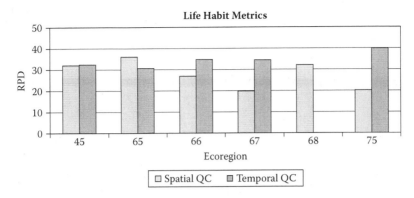

FIGURE 9.28 Comparison of relative percent difference (RPD) values averaged for all raw metric values of the life habit metrics per ecoregion designation and by quality control (QC) sample designation. (No temporal QC samples were collected for Ecoregion 68.)

measured are necessary for identifying the effects of measurement errors and inherent differences between sampling sites in relation to the overall variance of a metric or index on an ecoregional and subecoregional level (Karr and Chu 1999).

At first glance it is apparent that the majority of the average RPD values for the metric categories considered in this study (for both raw metric scores and standardized metric scores) are above the precision thresholds of the measurement quality objectives dictated by the QAPP document. The raw metric values for taxonomic richness and tolerance/intolerance metric categories appear to have better precision overall compared to community composition and life habit measures. For each ecoregion, the average RPD values for the taxonomic richness and tolerance/intolerance metric categories is nearest the QAPP-prescribed MQO value of 20%, while community composition and life habit measures are consistently above the 20% precision threshold.

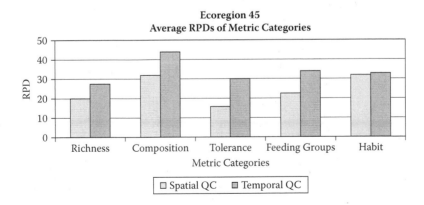

FIGURE 9.29 Comparison of relative percent difference (RPD) values averaged for all raw metric values by quality control (QC) sample designation per metric category for primary Ecoregion 45.

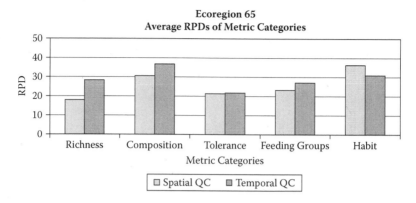

FIGURE 9.30   Comparison of relative percent difference (RPD) values averaged for all raw metric values by quality control (QC) sample designation per metric category for primary Ecoregion 65.

After standardization of the raw metric values, the average RPDs of the metric scores, again, were above the precision thresholds established by the QAPP document. While the average RPDs of the raw metric values for some metric categories were close to the prescribed precision threshold of 20%, the average RPDs for standardized metric scores for all of the metric categories are considerably higher than the precision criterion of 5%.

There appears to be some consistency among the averages of the metric RPDs from one ecoregion to another, as well as between the metric categories. With the exception of RPD averages for metrics associated with measures of tolerance/intolerance and taxonomic richness, which were at the 20% precision threshold, the remaining RPD averages were typically much greater than the prescribed MQO. It is evident that metrics falling into the richness and tolerance/intolerance categories generally have less variability and more precision than the other metric categories.

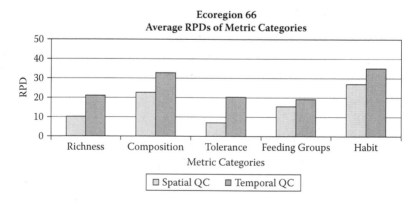

FIGURE 9.31   Comparison of relative percent difference (RPD) values averaged for all raw metric values by quality control (QC) sample designation per metric category for primary Ecoregion 66.

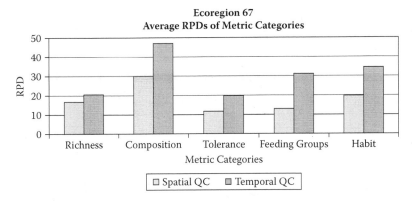

FIGURE 9.32 Comparison of relative percent difference (RPD) values averaged for all raw metric values by quality control (QC) sample designation per metric category for primary Ecoregion 67.

As stated earlier, RMSE levels were to be established as a result of this study. Ultimately, the ranges of RMSE considered to be acceptable for a bioassessment program will depend on the objectives of the water resource manager and the best professional judgment of the data analyst. The overall measures of error associated with biological data determine some level of data quality. Interpretations of data quality are important to the data user and decision makers to evaluate the degree of the reliance on technical and scientific information (Costanza et al. 1992). RMSE values are estimates of the standard deviation, which are also considered measures of precision. The assumption is that the larger the RMSE value, the less precision and greater variability within the measures.

Each metric category seemed to exhibit a "proportional" trend in variability when compared to one another, with the greatest ranges of variability still associated with the community composition measures. After metric value standardization, the range

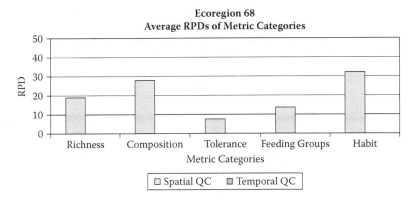

FIGURE 9.33 Comparison of relative percent difference (RPD) values averaged for all raw metric values by quality control (QC) sample designation per metric category for primary Ecoregion 68. (No temporal QC samples were collected for Ecoregion 68.)

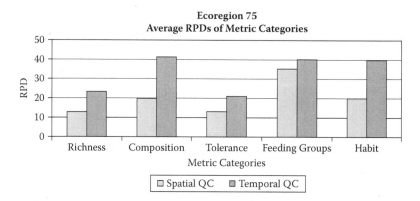

FIGURE 9.34 Comparison of relative percent difference (RPD) values averaged for all raw metric values by quality control (QC) sample designation per metric category for primary Ecoregion 67.

in variability for all metric categories increased significantly when compared to the RMSE values for the raw metric values. Additionally, the variability between each metric category generally became relatively uniform within each ecoregion.

For the precision measure CV, when compared to the RMSE values, an opposite trend in variability ranges occurred between raw metric values and standardized metric scores. This was most evident when comparing variability ranges of the community composition measures with CVs for the raw metric values, which were consistently almost twice the value of CVs for the standardized metric scores. CV value ranges for all other metric categories did decrease between raw metric values and standardized scores, but not as dramatically. Again, as with most other precision measures presented, community composition measures were still consistently much higher in their range of variability compared to the other metric categories.

Upon comparing the range of variability of RPDs, RMSEs, and CVs, it should be noted that these average values were inclusive only of the standardized metrics that were included in the final biotic index. Consequently, these average precision measures were based on a much smaller group of metrics (five to eight) than the other values presented, which were averaged for the entire suite of metrics. Additionally, the metrics chosen to represent the biotic index for an ecoregion were based upon their standardized values, primarily those metrics that provided the highest additive score and exhibited the best stress responses. Although the variability between the metrics considered to quantify the final bioassessment score was most likely minimal, resulting in smaller ranges of variability, the MQO criterion of 5% for RPDs between bioassessment scores was still not met.

When considering most of the raw count metrics (i.e., number of taxa per order, functional feeding groups, and habit), the values of RPDs are probably not as important as the evaluations of variance among the final bioassessment scores. Comparing the RPD values of the raw metric values and standardized metric scores to the RPD values for the final bioassessment scores, it is apparent that the range of overall variability decreased among the measures. Although the RPD values for the final bioassessment scores are still higher than the precision thresholds dictated by the QAPP for bioassessment scores, the values, overall, are much closer to the prescribed 5% threshold (as compared to the RPD values for raw metric values and standardized metric scores). As for the other precision measures of RMSE and CV, the range of values associated with the final bioassessment scores is also much narrower than the RMSEs and CVs calculated for raw metric values and standardized scores.

This initial examination of the analysis of data precision leads to two questions: (1) Assuming that SOP protocols for field sampling of invertebrates were followed with minimal error, what are the factors that could possibly influence the range of variability between an established sample site and its QC sample? (2) If we assume that the samples collected to determine the biological condition are a valid or characteristic representation of the ecosystem, then how should the data be interpreted in relationship to the predetermined threshold values for precision and data quality?

Considering the sampling methods implemented for bioassessment programs, there are many factors to consider after examining the RPD values for both raw metric values and standardized metric scores. The first factor to consider is the methodology for sampling the invertebrate community. As mentioned before, some bioassessment studies have centered on specific habitat types (i.e., riffles and/or runs) to determine the biological integrity of a lotic system. With the invertebrate sampling protocols used for this study, a multiple habitat sampling approach, while being more inclusive of the assortment of macroinvertebrates in freshwater systems, can also lend itself to greater variability because of the mixture of habitats sampled.

The "twenty-jab" method prescribed by the RBP protocol and the Georgia Department of Natural Resources (CSU 2000) was designed to sample a variety of habitats with relatively equal levels of effort in proportion to those habitats that typically occur in high- or low-gradient lotic systems. In instances where the designated 100 meter sampling reach did not provide the required distribution of effort among the different habitat types, the level of effort was "reallocated" and distributed evenly among those habitats that were more dominant in the sampling reach. This may be more of a concern when considering the replicability of spatial (200 meter) QC samples, if the variety of habitats designated for sampling did not occur equally from one sampling reach to the next. Therefore, some raw metric values may be distorted if jabs assigned to typically more productive habitats (i.e., riffles and snags) are replaced by jabs of less productive habitats (i.e., sand) and vice versa. These changes in the distribution of effort among habitat types can cause large variations in the invertebrate assemblages collected that can ultimately affect the range of metric values between sites and within all of the metric categories. In turn, this may be one factor responsible for higher and lower value RMSEs and CVs than may be expected, as well as RPDs above the precision thresholds predetermined for this project.

Another factor in the inclusion and exclusion of certain taxa from the composite samples collected is the use of random subsampling. Caton (1991) developed a gridded screen technique to increase objectivity in the laboratory subsampling of benthic macroinvertebrates. For biomonitoring programs, subsampling has been recommended as a valid and cost-effective procedure where time and monetary resources are limited. The rationale behind the use of subsampling is twofold, where the level of effort expended on each sample collected is relatively equal, and representative estimates of the invertebrate population sampled are selected or picked.

In some instances, it is possible for the sampling methodology to skew the average values of RPDs and RMSEs. In cases where there may be taxa that are rare, the occurrence of just one organism picked from either a primary or QC sample would cause an RPD value of 100% to be assigned to that metric if there were no occurrence of that same taxon in the corresponding QC or primary sample from the same stream. Although precision thresholds between samples would be high, the samples are most likely more similar than the RPD value may indicate. In these instances it may be necessary to consider the raw taxonomic data and their effect on the final value of the metric before assuming that the quality of the data is substandard.

Also, it is important to consider the characteristics of the metrics themselves and to what ecoregion or subecoregion they are being applied. For instance, the varying geomorphology across the state is directly responsible for the variability of habitats and water chemistry in the lotic systems being analyzed. This, in turn, will dictate the presence or absence of certain taxonomic groups and individual organisms based upon habitat requirements. Systems that are more dominated by high-gradient, headwater streams (e.g., Ecoregions 66 and 67) with allochthonous inputs will tend to have higher percentages of shredders and scrapers (e.g., Plecoptera and Coleoptera), whereas systems with low-gradient streams (e.g., Ecoregions 65 and 75) will tend to have higher numbers of filterers and collectors (e.g., Trichoptera and Diptera; Vannote et al. 1980). Therefore, some metric calculations and corresponding determinations of the average RPD, RMSE, and CV values for an ecoregion or subecoregion may not be truly indicative of the range of variability or quality of the data collected. Again, the data analyst will have to examine the individual sites within each subecoregion to assess the validity of the metric values for the region that was sampled. Fortunately, those candidate metrics that may not be significant or indicative of the invertebrate assemblages in an ecoregion or subecoregion are filtered out through DE calculations and box-and-whisker plot development during production of the biotic index.

The initial interpretation of RPDs for subecoregional sites must be considered cautiously. In instances where the RPD between two samples was calculated to be zero, there are two scenarios to consider: an RPD value of zero is the result of the raw numbers of the metric for the established sampling site and its QC to be either (1) equal in value or (2) for there to be no occurrence of the organisms that define the value of the metric in either sample. When examining the MQOs established for this project, a value of *zero* would appear to indicate that the original sample was a representative sample or there was minimal, or no error in performing the sampling (i.e., high data precision). However, the metrics themselves must be considered for their ecological significance to the target ecoregion. Since they are rare in that ecoregion, an RPD of zero for the number of Plecoptera in samples collected in the coastal plains (Ecoregion 75) is not as significant as RPD values for the abundant noninsect and Oligochaete taxa. The inclusion of certain metrics with minimal biological importance to an ecoregion or subecoregion can skew the overall average RPD values for those regions, as well as affect the ranges of RMSE and CV values.

For RPD values that were calculated to be 100, indicating absolute difference between QC and primary samples, some scrutiny is also deserved. An RPD value of

100 is essentially the result of a presence/absence scenario, where one sample may have as little as one individual, but the corresponding sample will have no occurrence of the same organism. Again, the presence of "rare" individuals from one sample compared to a corresponding QC or primary sample that may not have the same organism would not be as significant as the presence of 50 individuals where the corresponding sample may have none. The data analyst may need to examine RPD values for individual sites to determine if the presence or absence of certain organisms is significant in relation to the ecoregion being considered.

After standardization of raw metric values to produce metric scores, RPDs and RMSEs are, again, calculated to determine precision estimates of the collected data. Similarly, these values must also be examined with some scrutiny as the metrics are now ranked on a similar scale (i.e., values are expressed on a 0 to 100 scale) where some calculated values may be negative and others may be above the upper limit values of 100. Those calculated metric scores that fell into negative values or were above the upper limit values of 100 were changed to values of 0 or 100, respectively.

As outlined earlier, there are many steps in determining the appropriate suite of metrics that should be used to characterized and monitor the biological condition of an ecoregion. The metrics chosen are not only indicative of the aquatic assemblages of a certain region, but are also the most sensitive to anthropogenic stresses in the ecosystem. The RPD, RMSE, and CV values are based upon the standardized metric scores for those metrics included in the index. With the exception of a few metrics, the majority of the RPD values for the standardized metric scores of the metrics comprising each index are relatively high. Considering the precision thresholds dictated by the QAPP, the RPD values for standardized metric scores should ideally be less than or equal to 5%, but in relation to the overall trends that RPDs have exhibited for raw metric values and standardized scores, these results are consistent. Conversely, there appears to be no definitive correlation between the range or variability of the precision measures and the DE values associated with each metric chosen. All DEs for the metrics that formulated the final indices for the primary ecoregions did meet the minimal criteria of 50%, but higher DE values did not correspond with lower variability in the RPD, RMSE, or CV values for those metrics.

DE values at the subecoregional level improved, on average, when compared to the DE values at the primary ecoregional level. The metrics used to characterize the biological condition at the subecoregional level were more indicative of the differences between the reference and impaired condition at a smaller scale. Although DE values at the subecoregional level improved, there was no corresponding trend in the improvement (i.e., reduction in the range of variability) of the precision measures of RPD, RMSE, or CV for those metrics that constitute the biotic index.

When considering the ranges of variability between the metric categories and ecoregions and examining the average RPD, RMSE, and CV values for the metric categories, it must be noted that the number of QC sites among the ecoregions and subecoregions were not distributed evenly. This was because QC sites and those sites classified as reference or impaired (both spatially and temporally) were randomly selected. Therefore, final averages of the precision measurements for raw metric values and standardized metric scores may not have been weighted evenly. One example of this occurs for Ecoregion 68, which is comprised of only one subecoregion and had only one QC sampled for the impaired stream class. Upon examining the overall trends between the ecoregions, and for the majority of the precision measures applied, ecoregion 68 consistently exhibited a lower overall average RPD, RMSE, and CV when compared to other ecoregions. In instances where there was a minimal number of QC samples per ecoregion or subecoregion, the RPD, RMSE, and CV values associated with data precision and variability should be considered with caution, as the number of replicated samples may not be sufficient to illustrate ranges of variability within a certain ecoregion or subecoregion.

Similarly, the number of QC sites designated as reference/impaired and spatial/temporal was not distributed evenly among the stream class nor the QC sample type. Although not required by the QAPP for analysis, additional precision measures were considered to examine possible influences of variability between reference and impaired sites, as well as spatial and temporal variability by metric category across each primary ecoregion. The values consisted of overall averages of all metrics within each metric group per stream class and primary ecoregional designation. There does not appear to be any discernable overall pattern between metric category variability within each ecoregion (i.e., there is no consistency as to the range of variability between reference QCs when compared to impaired QCs). There are only a few exceptions to this, more specifically repeatable differences between stream classes in relation to metric categories and ecoregional designation.

One pattern that emerged was the difference in reference and impaired RPDs for the metric categories of taxonomic richness, functional feeding groups, and life habit from Ecoregion 75. The metric groups within Ecoregion 75 have the widest range of variability between the two stream classes, with the reference RPD values consistently being lower. Additionally, the range of variability for reference sites associated with Ecoregion 67 are, for all metric categories, is greater than the range for the impaired sites. In some respects, this particular pattern cannot be considered significant as the average impaired values was derived from only one sample.

Comparisons of RPD values for stream QC types were also considered at the primary ecoregional level by metric categories. It was apparent that there was consistency between the RPD values for spatial and temporal QC samples at both the ecoregional and metric category designation. With minimal exceptions, temporal QC sites had higher ranges of variability when compared to spatial QC sites. In general, this might be an expected conclusion if the sites that were originally chosen for sampling were indeed indicative of a "typical" reach of the catchment being analyzed. Also, considering that spatial QC samples were essentially collected at the same time, overall variability should be minimized.

When examining the average RPDs for the temporal QC samples, it was evident that at both the ecoregional level and by metric category designation that variability is much greater. Similarly, this might also be an expected result as there are many factors that can influence biological communities over time (i.e., rainfall patterns, temperature, and so forth). In trying to minimize the effects of temporal influences, a sampling "index period," was utilized for all samples collected. For sites that were designated for temporal QC sampling, field crews attempted to sample the phase QC stream at approximately the same time of year as previously collected.

Unfortunately, a precise determination of temporal affects on variability may not be identifiable from one sampling period to the next. One factor to consider is the number of "degree days" from year to year that cue the life stages of freshwater mac-roinvertebrates. Depending on daily temperature patterns between years, a sample collected one year may have third or fourth instar nymphs, which would, typically, be easier to identify to a lower taxonomic level. Other corresponding temporal QC samples may not have had the same number of degree days before sampling that could have resulted in earlier instar nymphs that may not be identifiable to the same taxonomic resolution as a previous sample. Ultimately, this will affect values of met-rics that require lower taxonomic resolution to be quantified.

More important, another temporal factor to consider is changes in land-use pat-terns across the state of Georgia. For some ecoregions (e.g., Ecoregion 45) that have highly urbanized areas (i.e., Atlanta), land use can change on a weekly basis. Considering that most land use data used to identify reference and impaired catch-ments are, at best, updated annually, the variability of the biological community from one year to the next could be extreme. This can also be exemplified in areas that have no urban influence at all, more specifically, areas with large amounts of acreage devoted to silviculture. Although less extreme in the rates of changes as compared to urbanized areas, it is not uncommon for mature stands to be clear cut by the time the same stream in the catchment is evaluated in the following year. Even with the required buffer strips emplaced to protect streams in these areas, there still exists the possibility of wide variances in the biological community from one sampling event to the next.

Apart from the nuances of the statistical analysis of this data, there are many biological factors that must also be considered when interpreting data precision and variability. As previously mentioned, some consideration must be given to individual metrics and their applicability to the ecoregion in question. In some instances, the ini-tial analysis of raw metric values may have not provided DE values at 50% or greater, which indicated the lack of differentiation between the biological character of a refer-ence and impaired stream. One such region that exemplified this was Ecoregion 75 (Southern Coastal Plains). The metrics that comprise the biotic index for Ecoregion 75 do not encompass measures from each metric category (i.e., taxonomic richness, func-tional feeding groups, etc.), but are limited to the measures of community composition and tolerant individuals. This is a direct reflection of the invertebrate assemblages that are most characteristic of the ecosystem in that region and are dictated, in part, by hab-itat features (i.e., predominance of sand and silt substrates, presence of woody debris, etc.). The majority of the organisms in the coastal plain ecoregion is noninsect (e.g. Amphipoda, Isopoda, Gastropoda, Oligochaeta, and so forth), being poorly described or quantified by traditional richness metrics (Gore et al. 2005).

Additionally, there were some metrics associated with the life habit category that were not considered in development of the biotic indices for all the ecore-gions. Specifically, the metrics for the life habit category that were excluded for use in the biotic indices included percentages of burrowers, climbers, sprawlers, and swimmers in the lotic community. Although the EDAS database did provide calculations for these metrics in question, there exists no definitive scientific litera-ture to support what type of stress response would be demonstrated by the organ-isms included in those groups, but rather their responses are inferred from other "lifestyle" characteristics (DeShon 1995; Fore et al. 1996; Barbour et al. 1999).

For example, those benthic macroinvertebrates whose characteristic life habit are classified as sprawlers, burrowers, and so forth are also categorized by other attributes such as a taxonomic order (i.e., Ephemeroptera, Plecoptera, and so on) and feeding mechanism (i.e., predator, shredder, collector, and so on). Therefore, stress responses that have been established at an order, tolerance, or feeding level have been correlated to a similar stress response for the organism in accord with its life habit characteristics.

In conjunction with the physical character of a habitat, the chemical character of the lotic ecosystem must be considered as a possible source of variability in macroinvertebrate communities. Many ecoregions and subecoregions contained both *clearwater* and *blackwater* streams. In contrast to clearwater streams, blackwater stream systems are typified by high tannin inputs (from terrestrial organic material), more acidic pH levels, and lower concentrations of dissolved oxygen. As a result, the benthic communities that dominate these systems can be significantly different from clearwater streams. The physical habitat of blackwater streams is characterized by sandy substrates and fine particulate organic matter, which serves as an ideal environment for oligochaetes, dipteran taxa, and mollusks, whereas clearwater streams in the same region would be dominated more by Trichoptera taxa and acid-intolerant Chironomid taxa (Meyer 1990). There was no separation of designated blackwater and clearwater streams in the analysis and it has been suggested that clearwater and blackwater streams may need to be categorized separately when developing biotic indices for a specific region, as these distinctive invertebrate communities may respond to anthropogenic stresses very differently (Gore et al. 2005).

Similarly, there has been discussion of establishing different suites of metrics for tidal streams in Ecoregion 75. In the initial selection process for selecting reference streams, there was no determination of the influences of tidal effects within coastal catchments. It became evident, after taxonomic identifications, that some samples contained invertebrate communities indicative of brackish water environments. Although these sites were not initially considered to be affected by estuarine influxes, the presence of salt-tolerant marine species, such as polychaetes and crabs, presented some problems in defining the biotic indices for streams influenced by tidal cycles. The primary problem is that the EDAS program does not account for metrics of marine species, which creates difficulty in developing biotic indices that are truly characteristic of the integrity of brackish water systems. To remedy this issue, Gore et al. (2005) suggested identification of a reference condition for both so-called *inland* streams and *tidal-coastal* streams when there is an occurrence of both systems in an ecoregion or subecoregion.

One major environmental factor that affected the majority of the southeastern portion of the state (i.e., south of the "fall line") was the occurrence of a sustained drought from 1999 to 2003 over most of the project's sampling phases. Many perennial streams in this region of the state were estimated to be dry for two years or more. This situation created two problems: (1) finding enough designated reference sites to satisfy the project requirements for the number of sites to characterize the biological condition, and (2) when designated reference sites did have water, in most cases, there was no initial indication that the sample collected would be representative of an "unstressed," typical biological community. Although Gore and Milner (1990) have demonstrated that disturbed lotic systems can be recolonized by macroinverte-brates in as little as 14 to 21 days, there was no sure way for field teams to verify that normal stream functions and invertebrate communities had returned to their typical character. In instances where lotic systems have had some form of sustained stress, additional sampling should be performed so that the reference and impaired condi-tions can be adequately defined.

With all of the possible influences of the biological variables discussed, ecologi-cal responses to varying levels and types of stressors can be complex and difficult to accurately measure with a high degree of reliability (Murtaugh 1996). In some cases, the use of benthic macroinvertebrates for bioassessment may not provide a clear response to anthropogenic influences. Floods and droughts inevitably will affect aquatic ecosystems over the course of time. Where there are instances of "pulse" events in an ecosystem, the induced stress may not be significant enough to perma-nently alter the composition of the aquatic community, especially invertebrates with their ability to recolonize (Gore and Milner 1990), fecundity, and dispersal ability (Patrick 1975). Pulse events can be either naturally occurring (i.e., droughts and floods) or human induced (i.e., industrial discharges).

In bioassessment programs, it is important to identify reference and impaired sites that encompass natural variability within and between watersheds, as well as the variability of the influences of possible anthropogenic influences. This is crucial to the subsequent calculated metrics and biotic indices that water resource managers will utilize in their decision-making process. Evaluation of stream ecosystem health can be hindered by the cost and time constraints posed by large-scale quantitative biomonitoring sampling protocols (Rosenberg and Resh 1993). The goal of quality control protocols is to measure the quality of a procedure so that it meets the needs of the user, while aiming to produce data that is dependable, adequate, and economi-cal (EPA 1995).

The concept of MQOs (also referred to as DQOs) in bioassessment programs is a useful tool in evaluating the consistency of data and limiting variability and potential sources of measurement error (Diamond et al. 1996). When comparing two samples to determine a level of precision, acceptable differences are typically predetermined by MQOs. These requirements for data quality should ideally be based on prior knowledge of sampling procedures and measurement variables that are specific to the region or ecosystem being studied (EPA 1989). Since there were no initial, wide-spread biological characterizations of the lotic invertebrate communities in the state of Georgia prior to this ecoregional study, the precision values of the MQOs stated for this project may have been unrealistic and unattainable for the ranges of the natural biological condition.

Considering the precision thresholds established for this study, there may be some questions or concerns about the validity, repeatability, and quality of the data used to determine the biotic indices for the ecoregions of Georgia. It is obvious that the majority of RPDs of the measurements parameters of metric values and standardized metric scores are above an established threshold, which is presumed to be indicative of some level of acceptable data quality. It is also evident that values for RMSE and CV are highly variable. The initial interpretation of these results may lead water resource managers to believe that the data are not very precise. In reality, these precision thresholds may have to be reevaluated and reestablished by additional sampling.

Because lotic invertebrate communities can vary significantly between geographic regions, it can be difficult, at first, to determine what should typify the reference or impaired biological condition of an ecoregion. In some uses of biomonitoring protocols, particularly the Index of Biotic Integrity (IBI) for fish communities, there is some expectation of what a fish community should exhibit under a reference (or least impacted) condition (Karr et al. 1986; EPA 1990). It is these reference expectations that can hinder their application to other geographic regions (Simon and Lyons 1995). Considering the invertebrate data produced by the Georgia Ecoregions Project, the consistency of all metric categories having average RPDs above the precision thresholds for both raw metric values and standardized metric scores may demonstrate that the lotic systems across the state of Georgia naturally have high variability from year to year and spatially within the catchment. This in turn may indicate that the established precision thresholds may not be indicative of the data quality for this specific project.

One possible method to determine if the variability within ecoregions and subecoregions is valid, or at least more indicative of the ecosystem, may be to standardize the number of replicate samples so that there is some equality in the level of effort expended for each subecoregion. For this study, a random subset of sites was chosen based on the total number of sites sampled throughout Georgia. As per the requirements of the QAPP, 10% of the primary sampling sites were chosen for QC sampling. There was no equal distribution of QC samples between or within subecoregions or ecoregions. In many scientific sampling protocols, the premise behind the use of an equal level of effort is to reduce bias and to improve consistency and repeatability (Plafkin et al. 1989; EPA 1995).

If the MQOs for this project, or any other bioassessment program, are not to be changed from the EPA (1995) guidelines, then there must be some evaluation of the rapid bioassessment protocols used and how it may be altered to achieve some established criteria for data quality. In a related research project, differences in subsample sizes were analyzed to determine if the prescribed 200 organisms to develop the biotic indices for the state of Georgia were adequate to characterize the necessary biological criteria. As we noted earlier, in most cases (for the ecoregions analyzed) a subsample of 300 organisms improved discrimination efficiencies, and better characterized the differences between the reference and impaired condition. With that preliminary research suggesting that the RBP methods used for Georgia may need to be altered, then it may be necessary to alter the bioassessment protocol further to achieve an acceptable level of data precision as generally mandated by the EPA.

With all of the biological and physical factors and variables previously discussed, more QC samples may need to be collected to better illustrate ranges of variability between raw metrics and final bioassessment scores. Especially in light of the comparisons between spatial and temporal QC samples, and in conjunction with the possible influences of a three-year drought, temporal variability may need to be addressed in more detail. As mentioned earlier, randomization of QC samples may not fully describe the ranges of variability, especially on a temporal scale. A more systematic QC sampling protocol should be employed for both reference and impaired stream classes, as well as from year to year.

Another factor to consider in relation to the variability among the temporal QC sites is climate patterns, primarily degree days from one ecoregion to the next. From a geographical and geomorphological context, Ecoregions 68, 67, and 66 (because of latitude and elevation) will typically have fewer degree days than Ecoregion 75 with a milder climate. As mentioned before, life stages could vary according to temperatures throughout the year, which, in turn could affect the emergence patterns of some macroinvertebrates. In colder climes, one index period may be sufficient to characterize the biological community, whereas in warmer regions, multiple index periods may be needed to ensure the collection of mutilvoltine invertebrate species.

The initial results from the four phases of this ecoregion project have been used to develop biocriteria specific for the ecoregions and subecoregions across the state of Georgia. The biocriteria used to develop the biotic indices must be reviewed through additional sampling in the future. Given the variability of hydrologic cycles over time, as well as changing land-use patterns as a result of urbanization or agricultural practices, biocriteria will not remain static. It is important to identify spatial and temporal variability in these aquatic systems so that biocriteria can be used wisely in water management decisions.

The specific objective of biomonitoring and bioassessment projects is to obtain the information needed to accomplish the project goals and uses. The ultimate goal is to characterize the biological condition of lotic ecosystems and determine which metrics adequately discriminate between levels of impairment, whether the impairment is minimal or severe. Biological metrics and biotic indices are used as a gauge of the biological condition, as well as being indicative of some type of response to anthropogenic stress. These measures of biological integrity are subject to change over the course of a prescribed study or continued monitoring. Many metrics may ultimately be revised and reevaluated for their effectiveness and applicability to the ecoregional character and the expectations of the water quality program in question.

REFERENCES

Allan, J. D. 1995. *Stream Ecology: Structure and Function of Running Waters*. New York: Chapman & Hall.

Barbour, M.T., J. Gerritsen, B.D. Snyder, and J.B. Stribling. 1999. *Rapid Bioassessment Protocols for Use in Streams and Wadeable Rivers: Periphyton, Benthic Macroinvertebrates and Fish*, 2nd ed. EPA 841-B-99-002. Washington, DC: U.S. Environmental Protection Agency, Office of Water.

Benke, A.C., R.L. Henry, III, D.M. Gillespie, and R.J. Hunter. 1985. Importance of snag habitat for animal production in Southeastern streams. *Fisheries* 10(5): 8–13.

Berger, W., H. McCarty, and R.K. Smith. 1996. *Environmental Laboratory Data Evaluation*. Amsterdam, New York: Genium Publishing Corporation.

Caton, L.W. 1991. Improved subsampling methods for the EPA "rapid bioassessment" benthic protocols. *Bulletin of the North American Benthological Society* 8: 317–319.

Clark, M.J.R., and P.H. Whitfield. 1994. Conflicting perspectives about detection limits and about the censoring of environmental data. *Water Resources Bulletin* 30(6): 1063–1079.

Columbus State University (CSU). 2000. *Quality Assurance Project Plan (QAPP): Ecoregions Reference Site Project for Wadeable Streams in Georgia.* Prepared for Georgia Environmental Protection Division and U.S. Environmental Protection Agency.

Costanza, R., S.O. Funtowicz, and J.R. Ravetz. 1992. Assessing and communicating data in policy-relevant research. *Environmental Management* 16(1): 121–131.

DeShon, J.E. 1995. Development and application of the invertebrate community index (ICI). *In Biological Assessment and Criteria: Tools for Water Resource Planning and Decision Making*, eds. W.S. Davis and T.P. Simon, 217–243. Boca Raton, FL: Lewis Publishers.

Diamond, J.M., M.T. Barbour, and J.B. Stribling. 1996. Characterizing and comparing bio-assessment methods and their results: A perspective. *Journal of the North American Benthological Society* 15(4): 713–727.

Feminella, J.W. 2000. Correspondence between stream macroinvertebrate assemblages and 4 ecoregions of the southeastern USA. *Journal of the North American Benthological Society* 19(3): 442–461.

Fore, L.S., J.R. Karr, and R.W. Wisseman. 1996. Assessing invertebrate responses to human activities: Evaluating alternative approaches. *Journal of the North American Benthological Society* 15(2): 212–231.

Gore, J.A., A. Middleton, D.L. Hughes, U. Rai, and P. Michele Brossett. 2005. *A Numerical Index of Health of Wadeable Streams in Georgia Using a Multimetric Index for Benthic Macroinvertebrates.* Atlanta: Georgia Department of Natural Resources.

Gore, J.A., and A.M. Milner. 1990. Island biogeographic theory: Can it be used to predict lotic recovery rates? *Environmental Management* 14: 737–753.

Hannaford, M.J., and V.R. Resh. 1995. Variability in macroinvertebrate rapid-bioassessment surveys and habitat assessments in a northern California stream. *Journal of the North American Benthological Society* 14(3): 430–439.

Hughes, R.M. 1995. Defining acceptable biological status by comparing with reference conditions. In *Biological Assessment and Criteria: Tools for Water Resource Planning and Decision Making*, eds. W.P. Davis and T.P. Simon, 31–47. Boca Raton, FL: CRC Press.

Hughes, R.M., and D.P. Larsen. 1988. Ecoregions: An approach to surface water protection. *Journal of the Water Pollution Control Federation* 60: 486–493.

Hynes, H.B.N. 1970. *The Ecology of Running Waters.* Toronto, Ontario: University of Toronto Press.

Intergovernmental Task Force on Monitoring Water Quality (ITFM). 1995. *The Strategy for Improving Water Quality Monitoring in the United States: Final Report of the Intergovernmental Task Force on Monitoring Water Quality.* Reston, VA: U.S. Geological Survey.

Karr, J.R., and E.W. Chu. 1999. *Restoring Life in Running Waters: Better Biological Monitoring.* Washington DC: Island Press.

Karr, J.R., K.D. Fausch, P.L. Angermeier, P.R. Yant, and I.J. Scholsser. 1986. *Assessing Biological Integrity in Running Waters: A Method and Its Rationale.* Special Publication 5, Illinois Natural History Survey, Champagne, IL.

Li, J., A. Herlihy, W. Gerth, P. Kaufman, S. Gregory, S. Urquhart, and D.P. Larsen. 2001. Variability in stream macroinvertebrates at multiple spatial scales. *Freshwater Biology* 46: 87–97.

Metzeling, L., and J. Miller. 2001. Evaluation of the sample size used for the rapid bioassessment of rivers using macroinvertebrates. *Hydrobiologia* 444: 159–170.

Meyer, J.L. 1990. A blackwater perspective on riverine ecosystems. *BioScience* 40: 643–651.

Mississippi Department of Environmental Quality (MDEQ). 2003. *Development and Application of the Mississippi-Benthic Index of Stream Quality (M-BISQ).* Jackson: Mississippi Department of Environmental Quality, Office of Pollution Control.

Murtaugh, P.A. 1996. The statistical evaluation of ecological indicators. *Ecological Applications* 6(1): 132–139.

Norris, R.H., E.P. McElravy, and V.H. Resh. 1993. The sampling problem. In *The Rivers Handbook*, Vol. 1, eds. P. Calow and G.E. Petts, 282–306. Oxford: Blackwell.

Omernik, J.M. 1995. Ecoregions: A spatial framework for environmental management. In *Biological Assessment and Criteria: Tools for Water Resource Planning and Decision Making,* eds. W.S. Davis and T.P. Simon, 49–62. Boca Raton, FL: Lewis Publishers.

Omernik, J.M., and A.L. Gallant. 1990. Defining regions for evaluating environmental resources. In *Global Natural Resource Monitoring and Assessments: Preparing for the 21st Century*, eds. H. G. Lund and G. Preto, 936–947. Bethesda, MD: American Society of Photogrammetry and Remote Sensing.

Patrick, R. 1975. Stream communities. In *Ecology and Evolution of Communities,* eds. M.L. Cody and J.M. Diamond, 445–459. Cambridge, MA: Belknap Press of Harvard University.

Plafkin, J.L., M.T. Barbour, K.D. Porter, S.K. Gross, and R.M. Hughes. 1989. *Rapid Bioassessment Protocols for Use in Streams and Rivers: Benthic Macroinvertebrates and Fish*. EPA 444-4-89-001. Washington, DC: USEPA, Office of Water.

Resh, V.H., and J.K. Jackson. 1993. Rapid assessment approaches to biomonitoring using benthic macroinvertebrates. In *Freshwater Biomonitoring and Benthic Macroinvertebrates*, eds. D.M. Rosenberg and V.H. Resh, 195–233. New York: Chapman & Hall.

Reynoldson, T.B., R.H. Norris, V.H. Resh, K.E. Day, and D.M. Rosenberg. 1997. The reference condition: A comparison of multimetric and multivariate approaches to assess water quality impairment using benthic macroinvertebrates. *Journal of the North American Benthological Society* 16(4): 833–852.

Roper, D.S. 1985. The role of biological surveys and surveillance. In *Biological Monitoring in Freshwaters: Proceedings of a Seminar, Hamilton, November 21–23, 1984,* Part I (Water and Soil Miscellaneous Publication No. 82), eds. R.D. Pridmore and A.B. Cooper, 17–20. Wellington, New Zealand: National Water and Soil Conservation Authority.

Rosenberg, D.M., and V.R. Resh, eds. 1993. *Freshwater Biomonitoring and Benthic Macroinvertebrates*. New York: Chapman & Hall.

Rosenberg, D.M., and N.B. Snow. 1977. A design for environmental impact studies with special reference to sedimentation in aquatic systems of the Mackenzie and Porcupine River drainages. In *Proceedings of the Circumpolar Conference on Northern Ecology, September 15–18, Ottawa*, Section III, 67–78. Ottawa, Ontario: National Research Council of Canada.

Stribling, J.B., and D.W. Bressler. 2004. *Data Quality and Performance Assessment: Phase 2 of the 303(d)/M-BISQ Project*. Prepared for the Water Quality Assessment Branch of Mississippi Department of Environmental Quality.

U.S. Environmental Protection Agency. 1989. *Preparing Perfect Project Plans: A Pocket Guide for the Preparation of Quality Assurance Project Plans*. EPA 600-9-89-087. Cincinnati, OH: USEPA, Office of Research and Development, Risk Reduction Engineering Laboratory.

U.S. Environmental Protection Agency. 1990. *Biological Criteria: National Program Guidance for Surface Waters*. EPA 440-5-90-004. Washington, DC: USEPA, Office of Water Regulations and Standards.

U.S. Environmental Protection Agency. 1995. *Generic Quality Assurance Project Plan Guidance for Programs Using Community-Level Biological Assessment in Streams and Wadeable Rivers*. EPA 841-B-95-004. Washington, DC: USEPA, Office of Water.

Vannote, R.L., G.W. Minshall, K.W. Cummins, J.R. Sedell, and C.E. Cushing. 1980. The river continuum concept. *Canadian Journal of Fisheries and Aquatic Sciences* 37: 130–137.

10 The Use of Rapid Bioassessment to Assess the Success of Stormwater Treatment Technologies (Best Management Practices) in Urban Streams

Erik Oij, James Banning, and James A. Gore

CONTENTS

In this chapter, the Rapid Bioassessment Protocol using criteria generated by the Georgia Ecoregions Project (Gore et al. 2005) was utilized to evaluate two types of best management practices (BMPs) located on two urban streams in the Columbus, Georgia, area. The two BMPs were built and designed to attenuate two very different issues to stream health: fecal coliform contamination and sediment loading.

BMPs represent a merge between effective pollution control options and practical management considerations. As defined by U.S. Department of Agriculture (1981), a BMP is a practice or combination of practices that are determined (by a state or designated area-wide planning agency) through problem assessment, examination of alternative practices, and appropriate public participation to be the most effective, practicable (including technological, economic, and institutional considerations) means of preventing or reducing the amount of pollution generated by nonpoint sources to a level compatible with water quality goals. Craddock and Hursh (1949) stated: "Today, better land-management practices must be inaugurated to restore a more favorable plant cover and soil structure if we wish to maintain land and stream conditions to serve our present and future needs for usable water." The better land management practices gave way to the BMPs of today.

Population growth will give rise to many continued problems for aquatic systems. In a period of 25 years (1990 to 2015), the number of cities with a population greater than 1 million people will increase from 118 to 272, and the number of megacities (>10 million people) from 14 to 27 (Shah 1996). Population growth will put a strain on a limited supply of potable water and increased pollution unless our tendencies and habits change dramatically. This growth will also increase urbanization and overall urban sprawl. Allan and Flecker (1993) identified six major causes of species loss in running waters. Three of the six causes can be attributed to urbanization: habitat loss and degradation, species invasion, and chemical and organic pollution. Stormwater runoff, untreated sewage, and industrial effluents are all major products arising from urban land uses. Sewage and industrial effluents have been widely studied and are further regulated as point sources of pollution. Urbanization exacerbates the effects of stormwater runoff on stream ecosystems. This is due to the coincident increase in impervious surfaces in the surrounding watershed. As the amount of impervious surfaces increase, the amount of area the water has to infiltrate the soil decreases. The reduction in area also limits the amount of water available for groundwater recharge, which in turns reduces stream base flow. In addition, urbanization forces more water from rain events to reach a stream faster, causing higher peak flows that lead to stream alteration and habitat degradation.

The reduction in stream base flow will alter the dynamic populations of the stream itself. With this reduction in flow, the availability of habitat will then also decrease over a period. Urban streams typically have a flashier hydrograph than rural streams. A typical hydrograph follows a gradual increase in flow and has a longer lag time than an urban stream. Urban streams' flow increases and falls rapidly over a short period of time. Along with the rapid increase in flow, urban streams are known for their elevated concentrations of nutrients and contaminants, altered channel morphology and stability, and reduced biotic richness (Meyer et al. 2005; Walsh, Fletcher et al. 2005; Walsh, Roy et al. 2005). This association with urban streams has been called the "urban stream syndrome" (Meyer et al. 2005; Walsh, Fletcher et al. 2005; Walsh, Roy et al. 2005).

Most urban streams have a very limited meander and are channelized to "protect" urban areas from flooding and to allow construction. With increased flows, intensity and frequency of scour are increased dramatically. The scour and the resulting fill are a function of the magnitude and distribution of sediment transported through a channel reach in time and space.

Channel morphology adjusts vertically and laterally (Niezgoda and Johnson 2005). The direction is imposed by the discharge and sediment loads through the process of erosion, transport, and deposition. These natural changes in channel structure have been limited by human intervention. Channelized urban streams are essentially flumes for water disposal, allowing the water to move at an increased velocity with little resistance. Sediment inputs to an urban stream system will also increase due to the removal of vegetation and a concurrent increase in impervious surfaces and stormwater discharge. Schueler (2000) analyzed the relationship between impervious cover and catchment conditions of hydrology, geomorphology, water quality, and biological condition into an urban stream classification model that can be used to predict existing and future quality of headwater streams. This system describes three primary classification categories based on percent of impervious cover: (1) sensitive (1% to 10%); (2) impacted (11% to 25%); and (3) nonsupporting (25% to 100%) (Niezgoda and Johnson 2005).

BMPs have been used to combat the many forms of degradation to urban streams. Typical BMPs include on-line or off-line filtration systems to remove organic particulates and bacteria, as well as sediment control methods off-line (buffer strips and barriers) or on-line (flow attenuation).

CASE STUDIES: APPLIED BMP PROJECTS

Weracoba Creek and Roaring Branch, two urban streams near the city of Columbus, Georgia, are both in Subecoregion 65c (Figure 10.1 through Figure 10.3 and Table 10.1). In Subecoregion 65c, the Sand Hills of Georgia form a narrow, rolling to hilly, highly dissected coastal plain belt stretching across the state from Augusta to Columbus. The region is composed primarily of Cretaceous and some Eocene-age marine sands and clays deposited over the crystalline and metamorphic rocks of the Piedmont (Ecoregion 45). Many of the droughty, low-nutrient soils formed in thick beds of sand, although soils in some areas contain more loamy and clayey horizons. On the drier sites, turkey oak and longleaf pine are dominant, while shortleaf-loblolly pine forests and other oak-pine forests are common throughout the region.

The next few sections will discuss the statistics and metrics from the Georgia Ecoregion Project that were applied to the evaluation of the BMPs located on Weracoba Creek and Roaring Branch. Table 10.1 lists the five characteristics that were measured while evaluating potential reference streams. The five metrics that were used to study the two streams are listed in Table 10.2 and their respective statistics are in Table 10.3.

In Subecoregion 65c there is slight overlap between reference and impaired conditions; however, the discrimination efficiency is near 100% (Figure 10.2) (Gore et al. 2005). The numerical rankings and resulting stream health conditions are described in Table 10.4 and Table 10.5.

65c – Sand Hills

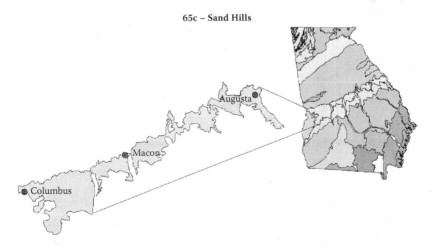

FIGURE 10.1 Map of the 65c—Sand Hills subecoregion in Georgia. Columbus, site of the BMPs, is located at the far Western border. (Adapted from Gore et al. 2005.)

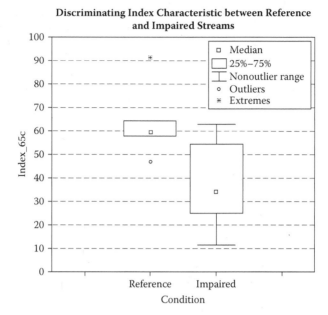

FIGURE 10.2   Discriminating index characteristic between reference and impaired streams for Subecoregion 65c—Sand Hills (Gore et al. 2005).

CASE STUDY 1: WERACOBA CREEK

Weracoba Creek is an urban stream in Columbus, Georgia, within the Lower Middle Chattahoochee River watershed (Bowman 2007). Weracoba Creek's watershed is approximately 10 square kilometers of predominately impervious surface (Figure 10.3). The creek was declared to be in violation of the Clean Water Act water quality standards, enforced by the state of Georgia. In particular the creek exceeds the total maximum daily loads for fecal coliform bacteria. This is an important consideration for the city of Columbus because Weracoba Creek feeds into the Chattahoochee River, which supplies the city with its drinking water.

The Columbus Water Works (CWW) has proposed a BMP to address the coliform loading problem within the watershed (see Figure 10.4). The BMP uses pretreatment in the form of an attenuation structure to reduce the large particles such as sand, oils, grease, and trash. The ultraviolet (UV) filter is operated continuously and filters both wet and dry weather events. During dry weather, the flow is attenuated and diverted through the UV filter only. The UV filter decreases bacterial populations in Weracoba Creek. As the flow increases, the flow passes through a compressed media filter prior to UV filter. With a further increase in flow, approximately one-third of the flow passes through the UV filter; the rest is either diverted through the compressed media filter or topples over the attenuation structure. The compressed media filter further removes debris and particulates.

To evaluate the effectiveness of this BMP, the Rapid Bioassessment Protocol was applied in accordance with GAEPD protocols and using the specific metrics

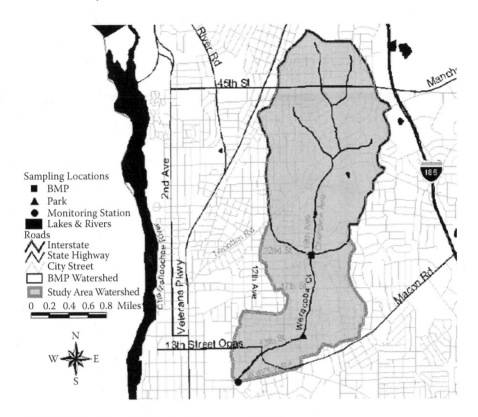

FIGURE 10.3 Weracoba Creek (Bowman et al. 2007).

prescribed for Subecoregion 65c (Gore et al. 2005). The pre-BMP condition of Weracoba Creek, sampled during the index period (September to March) prior to installation of the BMP was:

- Upstream of BMP site: Total metric score, 47.41; Category 3—Stream health rating (low) B
- Immediately downstream of BMP site: Total metric score: 37.80; Category 3—Stream health rating (low) B
- One kilometer downstream of BMP site: Total metric score: 54.75; Category 3—Stream health rating B

Thus, prior to installation of the BMP, Weracoba Creek, at best, could be rated as a Class B stream, indicating the need for continued monitoring and restoration efforts. The average index score of approximately 46 in the region of the proposed BMP classifies Weracoba Creek near the 50th percentile of the impaired condition for this subecoregion.

If the operations of the BMP were successful in improving water quality and habitat conditions, we expected to find an improvement in metric scores.

TABLE 10.1

Characteristic Reference Stream Land Use, Habitat, and Chemistry Data for Subecoregion 65c—Sand Hills

	Parameter	Mean	Median	Range
Catchment Land Use	% Natural	72.5	72.2	65.4–77.5
	% Agriculture	7.1	8.4	0–13.1
	% Silviculture	15.3	15.3	9.0–21.1
	% Urban	5.1	5.2	3.0–7.3
Habitat	Total Habitat Score (200)	164.4	164.0	159–170
	Epifaunal Substrate (20)	15.6	16.0	13–18
	Pool Substrate Characterization (20)	13.8	15.0	9–16
	Pool Variability (20)	14.8	16.0	10–16
	Sediment Deposition (20)	17.0	17.0	16–18
	Channel Flow Status (20)	19.0	19.0	19
	Channel Alteration (20)	18.4	19.0	17–19
	Channel Sinuosity (20)	11.8	13.0	9–15
	Bank Stability (L) (10)	8.8	9.0	8–9
	Bank Stability (R) (10)	9.2	9.0	8–10
	Vegetative Protection (L) (10)	8.4	8.0	8–9
	Vegetative Protection (R) (10)	8.4	8.0	8–9
	Riparian Vegetative Width (L) (10)	9.4	10.0	8–10
	Riparian Vegetative Width (R) (10)	9.8	10.0	9–10
In-Stream Habitat (substrate)	% Silt/Clay	37.0	12.0	0–22.8
	% Sand	87.0	95.7	63.0–100.0
	% Gravel	1.1	0	0–4.3
	% Cobble	0	0	0
	% Boulder	0	0	0
	% Bedrock	0	0	0
Chemistry (*in situ*)	Specific Conductivity (mS/cm)	0.020	0.015	0.003–0.049
	Dissolved Oxygen (mg/l)	11.3	11.7	10.3–12.5
	pH (SU)	5.1	5.1	4.3–6.2
	Turbidity (NTU)	2.3	1.1	0–6.9
Chemistry (laboratory)	Alkalinity (mg/l as $CaCO_3$)	1.8	0	0–8.2
	Total Hardness (mg/l as $CaCO_3$)	9.8	10.3	5.5–18.0
	Ammonia (mg/l as N)	0.054	0.052	BD–0.07
	Nitrate–Nitrite (mg/l as N)	0.18	0.11	0.07–0.47
	Total Phosphorous (mg/l as P)	BD	BD	BD
	Copper (mg/l)	BD	BD	BD
	Iron (mg/l)	0.54	0.54	BD–0.92
	Manganese (mg/l)	BD	BD	BD
	Zinc (mg/l)	BD	BD	BD

Source: Gore et al. (2005).
Note: BD = Below detection.

TABLE 10.2
Discriminating Invertebrate Metrics for Subecoregion 65c—Sand Hills Index 65c

Metric	Metric Category
% Trichoptera	Composition
Tolerant Taxa	Tolerance/Intolerance
Intolerant Taxa	
% Scraper	Functional Feeding Group
Clinger Taxa	Habit

Source: Gore et al. (2005).

TABLE 10.3
Descriptive Statistics for Reference Streams in Subecoregion 65c—Sand Hills

Metrics	DE	Minimum	Percentile (*n* = 5)					Maximum
			5th	25th	50th	75th	95th	
% Trichoptera	0.7	4.3	4.5	5.1	8.8	13.7	23.8	26.3
Tolerant Taxa	0.8	3.0	3.8	7.0	10.0	11.0	11.8	12.0
Intolerant Taxa	0.8	3.0	3.4	5.0	5.0	9.0	10.6	11.0
% Scraper	0.9	4.0	5.0	10.8	11.3	23.6	27.1	28.0
Clinger Taxa	0.6	10.0	10.2	11.0	12.0	15.0	16.6	17.0

Source: Gore et al. (2005).
Notes: *n* = Number of reference sites used; DE = discrimination efficiency between reference and impaired conditions.

TABLE 10.4
Description of Numeric Ranking for Subecoregion 65c—Sand Hills

Index Score	Numeric Ranking	Percentile (*n* = 15)
73 and above	1	Above 95th
61–72	2	Below 95th, above 75th
30–60	3	Below 75th, above 25th
20–29	4	Below 25th, above 5th
19 and below	5	Below 5th

Source: Gore et al. (2005).
Note: *n* = All reference and impaired sites in Subecoregion 65c.

TABLE 10.5
Stream Rating Based upon Numeric Ranking

Numeric Ranking	Stream Health Rating	Management Decision
1 or 2	A	Continue periodic monitoring to detect change baseline reference condition
3	B	Frequent monitoring critical to detect change in ecological status, lower range especially
4 or 5	C	Frequent monitoring necessary to determine remediation needs and if remediation has been successful

Source: Gore et al. (2005).

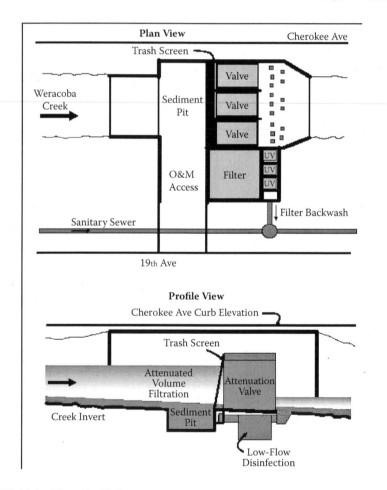

FIGURE 10.4   Weracoba BMP.

Using the specific metrics prescribed for Subecoregion 65c, we have determined that the post-BMP condition of the stream, sampled during the index period was:

- Upstream of BMP site: Total metric score: 53.42; Category 3—Stream health rating B
- Immediately downstream of BMP site: Total metric score: 61.61; Category 2 (marginal)—Stream Health rating A
- One kilometer downstream of BMP site: Total metric score: 58.33; Category 3 (marginal)—Stream health rating B

Although considerably improved, at the middle site, the metric scores evaluate the stream condition to be at the borderline between Class A and Class B streams, indicating that management strategies should include less frequent, but continued sampling, to be certain that the health of the stream is being sustained.

For the composite macroinvertebrate scores, there has been an improvement in each total score and changes within each individual score. It appears that the BMP has improved the biotic conditions in the stream (Table 10.6). This improvement is indicated by at least three of the five indicator metrics for the upstream site.

There is, apparently, no real change upstream of the BMP. However, downstream of the BMP, substantial improvements in the benthic macroinvertebrate communities are suggested.

TABLE 10.6

Comparisons of Metric Values at the Upstream Weracoba Creek Site before and after Implementation of the BMP

Metric Category	Metric	Predicted Response to Improved Stream Health	Weracoba Creek Upstream of the BMP Before	Weracoba Creek Upstream of the BMP After
Composition	% Trichoptera	Increase	5.08%; metric score = 10.62	1.90%; metric score = 3.97
Tolerance/ Intolerance	Tolerant Taxa	Decrease	6; metric score = 11.04	15; metric score = 18.40
	Intolerant Taxa	Increase	4; metric score = 14.72	3; metric score = 11.04
Functional Feeding Group	% Scrapers	Increase	3.95%; metric score = 6.43	6.64%; metric score = 10.81
Habitat	Clinger Taxa	Increase	3; metric score = 4.60	6; metric score = 9.20
		Total Metric Score	47.41	53.42
		Rank	3	3
		Grade	B	B

TABLE 10.7

Comparisons of Metric Values at the Middle Weracoba Creek Site before and after Implementation of the BMP

Metric Category	Metric	Predicted Response to Improved Stream Health	Weracoba Creek Upstream of the BMP Before	Weracoba Creek Upstream of the BMP After
Composition	% Trichoptera	Increase	4.29%; metric score = 8.97	25.73%; metric score = 18.40
Tolerance/ Intolerance	Tolerant Taxa	Decrease	7; metric score = 12.88	8; metric score = 14.72
	Intolerant Taxa	Increase	2; metric score = 7.36	5; metric score = 18.40
Functional Feeding Group	% Scrapers	Increase	2.45%; metric score = 3.99	2.43%; metric score = 3.96
Habitat	Clinger Taxa	Increase	3; metric score = 4.60	4; metric score = 6.13
		Total Metric Score	37.8	61.61
		Rank	3	2
		Grade	B	A

The site immediately downstream of the BMP shows an improvement in the total metric score (Table 10.7). This improvement is reflected by improvements in three of the five metrics used, the most significant change being the increase in the number of caddisflies (Trichoptera) in the sample. The percentage of Trichoptera (a composition metric) increased by over 20% (an almost sixfold increase). After placement and operation of the BMP for a year, there was an increase in the health index score to Category 2, the health rating advancing from B to (low) A.

At the site 1 kilometer downstream of the BMP, there has been an improvement in the biotic conditions in the stream, as indicated by at least two of the five metrics (Table 10.8).

At this lowest site sampled, the improvement in score is based upon the almost 60-fold increase in Trichoptera in the community. Trichoptera are considered to be indicators of improving stream health and this is reflected in the increase in total metric score.

Although these results suggest that there has been a significant improvement in the benthic macroinvertebrate communities downstream of the BMP, we suggest that a single monitoring event may not depict the true state of the health of the stream. Even those streams that are classified as minimally impaired (close to the reference condition)—Class A streams—must be continually monitored (albeit over a longer

TABLE 10.8
Comparisons of Metric Values at the Lower Weracoba Creek Site before and after Implementation of the BMP

Metric Category	Metric	Predicted Response to Improved Stream Health	Weracoba Creek Upstream of the BMP Before	Weracoba Creek Upstream of the BMP After
Composition	% Trichoptera	Increase	0.52%; metric score = 1.09	30.88%; metric score = 18.40
Tolerance/ Intolerance	Tolerant Taxa	Decrease	9; metric score = 16.56	7; metric score = 12.88
	Intolerant Taxa	Increase	3; metric score = 11.04	6; metric score = 18.40
Functional Feeding Group	% Scrapers	Increase	15.46%; metric score = 18.40	3.43%; metric score = 5.59
Habitat	Clinger Taxa	Increase	5; metric score = 7.67	2; metric score = 3.07
		Total Metric Score	54.75	58.33
		Rank	3	3
		Grade	B	B

interval, say, every two or three years) to assure that the stream health is being sustained. In this case, for example, the improved streams scored at marginal levels to be considered representative of Category 2 (A) streams. There is no doubt that significant improvement has been detected. However, the margin of analytical error easily places the stream condition score equally likely to be a high value in Category 3 as an even higher value in Category 2. Ultimately, the addition of new taxa (some will take a year or more to colonize) could result in even higher metric scores, moving the stream solidly into Category 2 or in the low values of Category 1; both indicating a stream health rating of A.

CASE STUDY 2: ROARING BRANCH

Roaring Branch, a small urban stream (USGS Hydrologic Unit Code [HUC] 03130003) in Columbus, Georgia, is located in the 65c Sand Hills subecoregion (Figure 10.2) (Gore et al. 2005). This stream is also very close to Subecoregion 45b (Southern Outer Piedmont), but after inspection, it was determined that Roaring Branch had the characteristics of Subecoregion 65c. This particular stream flows into Lake Oliver, a major source of potable water for the Columbus metropolitan area. This particular stream has the water quality problem of sediment loading. To alleviate this issue, a tributary that was determined to be a large source of the input of sediment had a BMP installed that

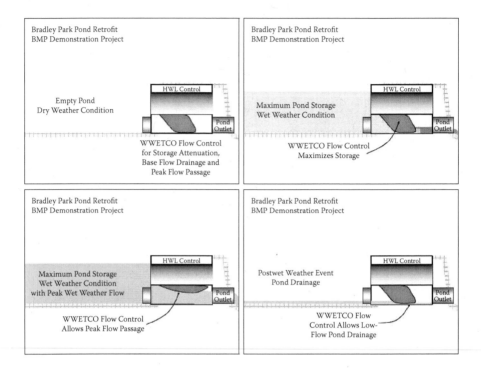

FIGURE 10.5   Roaring Branch BMP design (QAPP from WWETCO).

attenuated flow during precipitation events. The BMP (Figure 10.5) was installed in a small drainage pond adjacent to a shopping center. This drainage pond also received flow from a spring at its north end.

Sampling was conducted at a location just downstream of the BMP, on the tributary of Roaring Branch, and then downstream of the confluence of the tributary and Roaring Branch, itself (Figure 10.6). The sampling was conducted during the index recommended by the Georgia Ecoregions Project; that is, samples collected during the winter between the months of October through March (Gore et al. 2005). The initial sample was taken in December 2007, prior to the installation of the BMP. A summer sample was collected in August 2008, and the post-BMP winter sample was taken approximately one year later in October 2008.

Using the specific metrics prescribed for Subecoregion 65c, we have determined that the pre-BMP (Winter 2007) condition of the stream was:

- Immediately downstream of BMP (site "A") Winter 2007: Total metric score: 11.34; Category 5—Stream health rating, low C
- Roaring Branch (site "B") Winter 2007: Total metric score: 23.83; Category 4—Stream health rating, high C

Thus, prior to installation of the BMP, the tributary was classified as a very low Class C, indicating the need for frequent monitoring necessary to determine

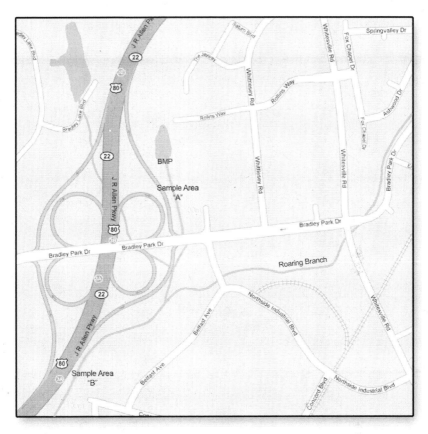

Roaring Branch – Muscogee County, Georgia

Sample Area "A" = Immediately downstream of BMP.
Sample Area "B" = Roaring Branch, downstream of confluence
1" = .175 miles.

FIGURE 10.6 Location of Roaring Branch sample sites.

remediation needs and, ultimately, if remediation has been successful. Roaring Branch, downstream of the tributary containing the BMP, was, at very best, a low Class B stream, which also needs frequent monitoring, critical to detect change in ecological status.

If the operations of the BMP had effectively improved water quality and habitat conditions, we expected to find an improvement in metric scores.

TABLE 10.9
Comparisons of Metric Values at the Tributary Site before and after Implementation of the BMP

Metric Category	Metric	Predicted Response to Improved Stream Health	Roaring Branch Just Downstream of the BMP Before	Roaring Branch Just Downstream of the BMP After
Composition	% Trichoptera	Increase	Metric score = 4.16	Metric score = 18.4
Tolerance/ Intolerance	Tolerant Taxa	Decrease	Metric score = 4.41	Metric score = 18.4
	Intolerant Taxa	Increase	Metric score = 0.88	Metric score = 18.4
Functional Feeding Group	% Scrapers	Increase	Metric score = 0.04	Metric score = 0
Habitat	Clinger Taxa	Increase	Metric score = 1.85	Metric score = 13.8
		Total Metric Score	11.34	69
		Numeric Ranking	5	2
		Stream Health Rating	C	A

Using the specific metrics prescribed for Subecoregion 65c, we determined that the post-BMP condition of the stream (Winter 2008–2009), sampled during the index period was:

- Immediately downstream of BMP (site "A") Winter 2008: Total metric score: 69; Category 2—Stream health rating, low A
- Roaring Branch (site "B") Winter 2008: Total metric score: 48.17; Category 1—Stream health rating, average B

Although considerably improved, the metric scores evaluated the stream condition to be at the borderline between Class A and Class B streams, indicating that management strategies should include less frequent, but continued sampling to be certain that the health of the stream is being sustained.

When examining the actual scores for each metric and for the total metric score for each site, there has been an improvement in each total score and changes within each individual score. It appears that the BMP has improved the biotic conditions in the stream (Table 10.9). At least four of the five indicators for the tributary site indicated improvement.

There is an apparent and significant change in habitat condition just downstream of the BMP. The population of macroinvertebrates increased at least 100-fold, as there were only 48 individuals in the entire composite sample that was collected prior to construction of the BMP (Winter 2007). In the Winter 2008 sample, 200 individuals were

TABLE 10.10

Comparisons of Metric Values Downstream of the Confluence of the BMP Tributary and Roaring Branch before and after Implementation of the BMP

Metric Category	Metric	Predicted Response to Improved Stream Health	Roaring Branch Just Downstream of the Tributary Before	Roaring Branch Just Downstream of the Tributary After
Composition	% Trichoptera	Increase	Metric score = 9.66	Metric score = 18.4
Tolerance/ Intolerance	Tolerant Taxa	Decrease	Metric score = 8.7	Metric score = 18.4
	Intolerant Taxa	Increase	Metric score = 3.86	Metric score = 3.68
Functional Feeding Group	% Scrapers	Increase	Metric score = 0	Metric score = 0
Habitat	Clinger Taxa	Increase	Metric score = 1.61	Metric score = 7.667
		Total Metric Score	23.83	48.17
		Numeric Ranking	4	3
		Stream Health Rating	C	B

collected from a small number of random subsampling squares, as opposed to sorting the entire sample collected prior to BMP installation just to obtain a small number of individuals. There was a large increase in Trichoptera (caddisflies), which are generally more intolerant to pollutants and other impairments to flow.

Immediately downstream of the confluence of the tributary and Roaring Branch, there was a slight improvement in the total metric score (Table 10.10). This improvement is reflected by changes in four of the five metrics assessed, the most significant change being an increase in the number of individuals in the sample. Again, similar to the tributary sample, the entire subsampling tray was processed resulting in only 105 individuals pre-BMP. In the post-BMP sample, more than 200 individuals were extracted from approximately 7 random squares out of the 30 in a 1 square meter subsample. This increase alone shows an improvement in stream health, as indicated by macroinvertebrate community composition. After placement and operation of the BMP for a year, there was an increase in the health index score to a low Category 1, the health rating advancing from *B* to *A*.

The Georgia Ecoregions Project's metrics for invertebrate populations are based upon an index derived from winter samples, but we also chose to collect a summer sample. There was only one summer sample taken, in September 2008, approximately nine months after the BMP's installation. The results from this summer sample also support the conclusion of an improvement in water quality (Table 10.11).

TABLE 10.11

Comparisons of Metric Values at the Tributary and Confluence Sites on Roaring Branch after Implementation of the BMP (Summer 2008)

Metric Category	Metric	Predicted Response to Improved Stream Health	Roaring Branch Tributary Just Downstream Summer 2008 After	Roaring Branch Just Downstream of the Tributary Summer 2008 After
Composition	% Trichoptera	Increase	Metric score = 18.4	Metric score = 18.4
Tolerance/ Intolerance	Tolerant Taxa	Decrease	Metric score = 18.4	Metric score = 18.4
	Intolerant Taxa	Increase	Metric score = 11.04	Metric score = 3.68
Functional Feeding Group	% Scrapers	Increase	Metric score = 12.52	Metric Score = 13.56
Habitat	Clinger Taxa	Increase	Metric score = 13.8	Metric score = 10.73
		Total Metric Score	74.16	64.77
		Numeric Ranking	1	2
		Stream Health Rating	A	A

Even though some of the individual species assemblages were very different from the winter sample, there seemed to be a significant improvement in stream health, as indicated by macroinvertebrate taxa in the summer sample.

Although results from both case studies suggest that there has been a significant improvement in the benthic macroinvertebrate communities downstream of the BMP, we suggest that a single monitoring event may not depict the true state of the health of the stream. Even those streams that are classified as minimally impaired (close to the reference condition)—Class A streams—must be continually monitored (albeit over a longer interval) to ensure that stream health is being sustained. In this case, for example, the improved streams scored at marginal levels to be considered representative of Category 2 (A) streams. There is no doubt that significant improvement has been detected. However, the margin of analytical error easily places the stream condition score equally likely to be a high value in Category 3 as a higher value in Category 2. Ultimately, the addition of new taxa (some will take a year or more to colonize) could result in even higher metric scores, moving the stream solidly into Category 2 or in the low values of Category 1; both indicating a stream health rating of A.

REFERENCES

Allan, J.D., and A.S. Flecker. 1993. Biodiversity conservation in running waters. *Bioscience* 43: 32–43.

Bowman, J. 2007. *Demonstration of Disinfection and Stormwater Pollutant Load Reduction BMP in Weracoba Creek Watershed.* Prepared for Columbus Water Works (CWW). Available from: Wet Weather Engineering and Technology, 800 Lambert Drive, Suite F, Atlanta, GA 30324.

Craddock, G.W., and C.R. Hursh. 1949. Watersheds and how to care for them. In *Trees: The Yearbook of Agriculture*, 603–609. Washington, DC: U.S. Government Printing Office.

Gore, J.A., A. Middleton, D.L. Hughes, U. Rai, and P. Michele Brossett. 2005. *A Numerical Index of Health of Wadeable Streams in Georgia Using a Multimetric Index for Benthic Macroinvertebrates.* Atlanta: Georgia Department of Natural Resources.

Meyer, J.L., M.J. Paul, and W.K. Taulbee. 2005. Stream ecosystem function in urbanizing landscapes. *Journal of the North American Benthological Society* 24: 602–612.

Niezgoda, S.L., and P.A. Johnson. 2005. Improving the urban stream restoration effort: identifying critical form and processes relationship. *Environmental Management* 35: 579–592.

Schueler, T.R. 2000. The importance of imperviousness. In *The Practice of Watershed Protection: Techniques for Protecting Our Nation's Streams, Lakes, Rivers, and Estuaries*, ed. T.R. Schueler, 7–18. Elliott City, MD: The Center for Watershed Protection.

Shah, A.A. 1996. Urban trends and the emergence of the megacity. In *The Future of Asian Cities: Report of 1996 Annual Meeting on Urban Management and Finance*, ed. J.R. Stubbs, 11–31. Manila, Philippines: Asian Development Bank.

U.S. Department of Agriculture, Soil Conservation Service. 1981. *Land Resource Regions and Major Land Resource Areas of the United States* (Agricultural Handbook 296). Washington, DC: U.S. Government Printing Office.

Walsh, C.J., T.D. Fletcher, and A.R. Ladson. 2005. Stream restoration in urban catchments through redesigning stormwater systems: Looking to the catchment to save the stream. *Journal of the North American Benthological Society* 24: 690–705.

Walsh, C.J., A.H. Roy, J.W. Feminella, P.D. Cottingham, P.M. Groffman, and R.P. Morgan, II. 2005. The urban stream syndrome: Current knowledge and the search for a cure. *Journal of the North American Benthological Society* 24: 706–723.

11 Implementation of the Rapid Bioassessment Protocol

Michele P. Brossett, Duncan L. Hughes,
Michele de la Rosa, and James A. Gore

CONTENTS

Regulatory agencies use various bioassessment methods to determine stream quality. Stream health is often classified as meeting attainment or nonattainment of biological integrity; thus, biocriteria are the benchmarks for water resource protection and management (Gibson et al. 1996). Once steam health is determined, criteria developed, and the bioassessment method tested, the next step for the regulatory agency is to determine how the criteria will be implemented. There are different methods, procedures, and regulations each regulatory agency must follow, but there is some common ground about how and why states implement the biocriteria.

WATER QUALITY CRITERIA AND STANDARDS

According to Section 303(c) of the Clean Water Act (CWA), the regulation of a water quality standards program is envisioned as a joint effort between states and the United States Environmental Protection Agency (EPA; Barbour et al. 1999). State regulatory agencies are responsible for setting, reviewing, revising, and enforcing water quality standards. The EPA's responsibilities include development of regulations and policies, as well as guidance in implementation. The EPA has the authority to review and either approve or disapprove a regulatory agency's standards and, when necessary, to create federal water quality standards.

Criteria development is important in order to protect the uses of a water body while preventing degradation of water quality. Standards are adopted to protect the public health or welfare, enhance the quality of water, and to protect biological integrity. Although not strictly oriented toward the protection of public health and welfare, protection of biological integrity is required to be included as part of each regulatory agency's water quality standards, per Sections 303 and 304 of the CWA (Barbour et al. 1999). One way to accomplish the goal of preservation of integrity is through the development and use of biological criteria (biocriteria). Biological criteria can provide scientifically sound and detailed descriptions of the designated aquatic life use for a specific water body or segment. Biocriteria establish the biological benchmarks for directly measuring the condition of aquatic biota and determining water quality goals, as well as setting management and monitoring priorities and evaluating the effectiveness of implemented controls and management actions.

DESIGNATED USE

Drinking water supply, primary contact recreation (e.g., swimming), and aquatic life support are some of the primary examples of beneficial uses designated by most regulatory agencies. Designated uses each have unique water quality requirements that must be met in order for the designated use to be supported. Water bodies may have multiple beneficial uses. When a water body does not attain the water quality standards needed to support its designated use, the water body is considered to be impaired.

Subcategories are used to refine and clarify designated use classes when several surface waters with distinct characteristics fit the same use class or when waters do not fit well into a single category (Barbour et al. 1999). Biosurvey analysis may reveal unique differences between aquatic communities that inhabit different waters with the same designated use. Measurable biological attributes can be used to refine aquatic life use or can separate classes of aquatic life into two or more subcategories. Ohio, for example, established the exceptional warm water use class to include all unique waters.

BIOCRITERIA

Biological criteria provide an evaluation standard for direct assessment of the conditions of the biota that live either part or all of their lives in aquatic systems (Gibson et al. 1996), by describing, in narrative or numeric form, the expected biological condition of a minimally impaired or unimpaired aquatic ecosystem. Biocriteria can be used to define ecosystem rehabilitation goals and assessment endpoints, and supplement traditional physicochemical measurements, as a refinement to evaluating nonpoint source pollution, as well as physical and biological stressors. Biological criteria can also be used by regulatory agencies to refine the aquatic life use classifications including protection and propagation of fish, shellfish, and wildlife.

Well-written and effective biocriteria provide scientifically sound evaluations that protect the most sensitive biota and habitats; protect healthy, natural aquatic communities; support and strive for protection of chemical, physical, and biological integrity; include specific assemblage characteristics required for attainment of designated uses; are clearly written and easily understood; adhere to the philosophy and policy on nondegradation of water resource quality; and are defensible in a court of law (Gibson et al. 1996). Biocriteria should be set at levels sensitive to anthropogenic impacts that pristine or minimally impaired sites are in danger of being classified out of attainment and robust enough that severely impaired sites do not receive certificates of health or attainment. Biocriteria that closely represent the natural biota will protect against further degradation and stimulate restoration of degraded sites.

Water resource managers can choose to use numeric or narrative criteria for evaluating stream conditions. Only a few regulatory agencies in the United States have adopted narrative biocriteria, and only three have adopted numeric biocriteria (Yoder and Barbour 2009). Most federal agencies prefer that regulatory agencies adopt refined designated aquatic life uses and numeric biocriteria as part of their water quality standards, arguing that impaired waters would be given more accurate identifications and descriptions, leading to better planning and management decisions.

NARRATIVE CRITERIA

North Carolina and Maine have adopted narrative biocriteria, which are based upon the assumption that the original structure and function must be maintained. Maine's biocriteria, "as it naturally occurs," are defined by specific ecological attributes, such as taxonomic equality, numeric equality, and the presence of specific pollution tolerant or intolerant species. Similarly, point and nonpoint source pollution are evaluated by North Carolina using narrative criteria that define acceptable levels of taxonomic richness, biotic indices of community functions, and numbers of individuals of key species. In combination, these narrative criteria determine the ratings for designated use as poor, fair, good-fair, good, and excellent. However, the state of Georgia has narrative criteria for the fish Index of Biotic Integrity (IBI), ranking streams as excellent, good, fair, poor, or very poor, but has not adopted these criteria, instead assessing physicochemical levels as compliance with CWA regulations.

Narrative biological criteria are not accepted without a quantitative database to support them (Gibson et al. 1996). Implementation of quantitatively based narrative biocriteria allows for accumulation of monitoring data and continual improvement in analysis of indicators and classification systems; thus, they can be adjusted through the administrative process rather than amending state laws.

NUMERIC CRITERIA

Numeric biocriteria are described by a defined range of values, not a single number, and attempt to account for natural variability in a healthy environment (Gibson et al. 1996). Numeric biocriteria include discrete quantitative values that summarize the status of the biological community, thereby describing the expected condition for systems with different designated uses. The numeric criterion allows a level of specification for water resource evaluations and regulations not common to narrative criteria.

Resident biota are sampled at minimally impaired sites to establish reference conditions (Gibson et al. 1996). Such values as species richness, presence or absence of indicator taxa, and distribution of trophic groups establish the normal range of conditions within the biological community as they exist in unimpaired systems.

For numeric biocriteria, careful assessments of biota in multiple reference sites are required to establish the most broadly applicable criteria. The multimetric approach, which incorporates species richness, trophic composition, abundance or biomass, and organism condition, maximizes the ability to detect subtle changes in community structure and function in response to subtle changes in physical or chemical condition. For example, regulatory agencies in Ohio combine narrative and numeric criteria for evaluation of fish and macroinvertebrates in streams and rivers. Fish communities are assigned numeric criteria based upon the Index of Well-Being (IWB) and the Index of Biotic Integrity (IBI). The macroinvertebrate community is assessed with Ohio's Invertebrate Community Index (ICI) (Ohio EPA 1999). The minimum warm water habitat criteria for each index is set at the 25th percentile, recorded from the established reference sites within the ecoregion. The exceptional warm water habitats are set at the 75th percentile, as evaluated from the statewide set of reference sites.

APPLICATIONS OF BIOCRITERIA

Currently, the EPA does not require the use of biocriteria as regulatory limits in National Pollutant Discharge Elimination System (NPDES) permits (Gibson et al. 1996). The EPA does recommend all state regulatory agencies develop and use biocriteria as a permit assessment tool. States are encouraged to evaluate the success of pollution control efforts using biocriteria. The nonpoint source program, in some states, uses biocriteria to determine the success of restoration projects. Some state regulatory agencies now require watershed assessments and watershed protection plans for biological, chemical, and physical parameters for wastewater permit holders. Counties, cities, and municipalities continue to monitor steams on a long-term basis and implement best management practices (BMPs) to improve biocriteria.

Because of limited funds, manpower, resources, and time, management planning is essential. Regulatory agencies can rank and combine similar water bodies as a way to assign priorities (Gibson et al. 1996) and then can focus on water bodies that are most likely to respond to restoration and those that are likely to not attain their designated use. State programs (i.e., implementation of CWA Section 319(h) and total maximum daily load [TMDL]; Section 303(d)) help fund many types of restoration projects. As demonstrated by the evaluations of BMPs in Chapter 10, regulatory agencies are able to determine which projects should continue or need further modification.

As more biological data are collected for the creation and modification of biocriteria, regulatory agencies will be able to enhance their knowledge of water quality and stream integrity. Currently, chemical, physical, and biological criteria are applied independently in a regulatory context. The criterion for one parameter does not influence the application of other criteria. We suggest that regulatory agencies continue to investigate the usefulness of the application of weights to the various criteria, in combination, as an alternative approach to evaluating water quality.

VALIDATION AND IMPLEMENTATION OF BIOCRITERIA

After biocriteria are developed, the next steps are validation of reference condition and survey techniques and the implementation of the program at various sites within watersheds with subsequent determinations of impairment (Gibson et al. 1996). For the most efficient validation, regulatory agencies need to continue to evaluate and revise sampling and evaluation criteria as needed. Biological ecosystems are dynamic systems, thus changes in biocriteria should be expected as ecosystems evolve. Chemical criteria tend to be more static under current regulatory definitions. However, biocriteria can be revised when better information is available, natural conditions have changed, or the waters of interest have improved (Gibson et al. 1996). The validation process comprises statistical analysis of biological, physical, and chemical data to establish natural variability and validity of the existing biocriteria. The classification of stream conditions should be continuously adjusted if biological and geographical data support changes. For example, the data collected for the study in Georgia classified stream conditions based upon an ecoregional and subecoregional approach; however, the EPA has suggested a newer classification of streams should instead be based upon bioregions (Michael Barbour, personal communication, May 2004).

Combining other data sources with biological assessment, water resource managers can accomplish a more comprehensive evaluation (Gibson et al. 1996). The source of impairment must be identified prior to remedial action. Beyond point sources of contamination, the most common nonpoint sources of impairment are likely to be alteration of habitat, change flow characteristics, altered biological interactions, or increases in runoff (with associated changes in sedimentation or nutrient concentrations). Once probable causes are identified, remediation can begin and continual monitoring can determine the progress of recovery.

DISCUSSION AND CONCLUSIONS

The Rapid Bioassessment Protocols (RBPs), using a macroinvertebrate metric index that is specific to each ecoregion and subecoregion, have been shown to be an effective mechanism to assess stream health and to prioritize monitoring and restoration activities within each ecoregion or subecoregion (Gore et al. 2005, 2006). Although it may be necessary to modify some of those procedures to ensure greater confidence in the index values generated (such as increasing the number of individuals removed from each subsample and some amount of *a priori* examinations of sampling areas when reallocation is necessary; see Chapter 7), the rapid bioassessment procedure provides a large database of macroinvertebrate distributions and tolerance metrics, as well as a dynamic procedure for creating effective points of comparison, the "reference condition." The production of a prioritized listing of activities related to overall stream health is compliant with requirements within Section 319 of the Clean

Water Act and provides a basis for application to evaluation of TMDLs, as required by Section 303(d).

Many state regulatory agencies (see, for example, Ohio EPA 1999) have already established an objective of integrating program activities around the TMDL program and a watershed-based approach to assessments of wadeable streams. The establishment of a rotational five-year basin approach allows better coordination of the collection of monitoring data so that information will be readily available for revision and enforcement of new water quality standards. Within each basin, sampling surveys should accomplish the following objectives:

1. Provision of current water quality conditions, especially in those catchments scheduled for TMDLs in the near future.
2. Determine the extent to which streams in each catchment have attained the stream health standards, as currently defined by the metrics provided by the rapid bioassessment program (Gore et al. 2005).
3. A determination if the macroinvertebrate metrics assigned to a given stream are appropriate to the ecoregion or subecoregion or if they need to be modified, based upon current sampling results.
4. A determination if changes in existing biological, chemical, or physical indicators have taken place, as a result of implementation of point (NPDES) or nonpoint source pollution (BMP) controls and the resulting improvements in the bioassessment index for that stream since implementation.

The findings and conclusions of each regional survey, as well as those accomplished in key basins, may factor into regulatory actions taken by regulatory agencies and should be incorporated into water quality permits, water quality management plans, and water quality reports prepared to meet the requirements of Sections 305(b) and 303(d) of the Clean Water Act.

Section 305(b) of the Clean Water Act requires the state to report how well the waters of the state support the beneficial uses of fishing, aquatic life, and recreation. As described by Gore et al. (2006), a priority listing of stream health can be achieved through application of the RBP process (Chapter 6) and could be directly applied to the category of aquatic life. With the addition of the IBI component of fisheries evaluation (see Barbour et al. 1999), the RBP can also address the value of the streams for fishing. For example, the state of Ohio lists it streams as:

Part I—Waters that are impaired or threatened by one or more pollutants
Part II—Waters that are impaired or threatened by pollution only (habitat impairment)
Part III—Waters that have had a TMDL established and approved but where standards have not yet been met
Part IV—Waters that are currently impaired, but where technology improvements are expected to result in attainment of impaired waters by the next TMDL cycle

In these cases, the recommended classification presented in Chapter 6 (see also Gore et al. 2005, 2006) allows the comparison of all streams to a reference condition and those streams found to be impaired can be further reclassified to meet established goals and implementation of point or nonpoint source controls.

Once Section 305(b) listing is accomplished, a typical TMDL process also requires an assessment of water bodies to determine degree of impairment by designated use and then to determine the causes and sources of impairment to complete the TMDL support documentation. The RBPs described in this book allow a regulatory agency to classify and prioritize stream health as the initial steps in Section 305(b) and TMDL processes. Indeed, because the reference conditions are generally based upon ecoregional differences in land use, there is some ability to identify potential sources and causes of impairment as part of the TMDL process.

The next steps in the TMDL process are to identify target conditions in any catchment of concern. This includes the analysis of existing data and determination of geographical and social/cultural concerns among the stakeholders in that ecoregion or subecoregion. Although the bioassessment data cannot aid in stakeholder identification, it can certainly provide information on reference conditions specific to that region and a classification of the streams (based upon a 5-point scale) in that catchment so that stakeholders can understand the priorities that must be addressed.

In the next phase of the TMDL process, stakeholders and the regulatory agency must work together to develop restoration targets. This process will involve identification of the existing load, the desired load to achieve the restored condition, and developing restoration scenarios based upon the restoration target; that is, the achievement of the reference conditions (either 1A or 2A streams).

After selection of an approved restoration scenario and appropriate mileposts for success (that is, increasing levels of macroinvertebrate metric performance), the regulatory agency and the stakeholders can jointly implement a restoration plan and schedule. The restoration should include regular sampling of the streams, both before and after implementation of point-source controls or BMPs for nonpoint sources, so that continuous comparisons can be made with existing (or subsequently modified) reference conditions for that region. At the point that the chemical water quality status and the stream health (as measured by RBP metrics for that ecoregion or subecoregion) have been verified, the water body can be delisted or relisted as appropriate.

Federal and state regulatory agencies are examining the processes to integrate the results of biological, habitat, chemical, and toxicological assessments in making the determination of an aquatic life use support. One method that the EPA suggests is the biological condition gradient and the tiered aquatic life uses (TALU). The Causal Analysis/Diagnosis Decision Information System (CADDIS) is one EPA tool to aid in the identification of water bodies that are not meeting their designated use by comparing established biocriteria.

TALU uses refined tools, criteria, and broader support for all relevant water quality management programs (Yoder and Barbour 2009). A TALU-based approach delivers support at the planning stage of the management process by assuring water quality standards are both appropriate and attainable prior to development of abatement strategies and responses. TALU ensures that management efforts are targeting greatest need and value.

CADDIS is an online application that helps find, access, organize, use, and share information to conduct causal evaluations in aquatic systems. Based upon the EPA's stressor identification process, CADDIS is a formal method for identifying causes of impairment. CADDIS can be a lengthy and time-consuming process. However, when regulatory agencies are unable to determine the cause of impairment by best professional judgment or in special studies, CADDIS will be an invaluable tool.

Measurement quality objects (MQOs) need to be completed for both the field (spatial, duplicate reach quality control sample; and temporal, phase quality control sample; as discussed in Chapter 9 and Stribling et al. 2008) and laboratory. Without MQOs, the quality of the data is uncertain. Each regulatory agency, water resource manager, or water quality program must establish the MQOs to assure that data quality is acceptable or whether corrective actions are warranted.

It would appear that the adoption of narrative criteria in standards and regulations is appropriate and, commonly, categories such as very good (or excellent), good, fair, poor, and very poor, associated with numeric criteria, have been employed by regulatory agencies. Numeric criteria will change over time as more data are available and as streams degrade or improve. The categories can remain constant while underlying numerical descriptors change. These narrative designations, then, preclude the need to require formal legislation to approve new designations or definitions. Despite these broad narrative categories, regulatory agencies remain divided on the attainment of designated use for those streams that are classified at the median value (i.e., fair). Alabama and Florida, for example, list fair streams as impaired, not meeting designated uses. However, North Carolina lists fair streams as supporting their designated uses. Tennessee initially places fair streams in a category requiring additional data. Resampling efforts determine attainment of designated use. If the stream is reevaluated as better than fair it is classified as " in attainment" or "not in attainment," if reevaluation describes a fair or worse condition.

Despite the minor controversies over classification of the median condition, a multimetric design (like the RBP) or a multivariate design are the most effective means to incorporate biological response variables and changes in ecosystem structure into the process of meeting water quality objectives. We suggest that regulatory agencies evaluate both techniques as a means of better managing a resource that is critical to the survival of aquatic life and the integrity of our running water resources.

REFERENCES

Barbour, M.T., J. Gerritsen, B.D. Snyder, and J.B. Stribling. 1999. *Rapid Bioassessment Protocols for Use in Streams and Wadeable Rivers: Periphyton, Benthic Macroinvertebrates and Fish*, 2nd edition. EPA 841-B-99-002. Washington, DC: U.S. Environmental Protection Agency, Office of Water.

Gibson, G.R., Jr., M.T. Barbour, J.B. Stribling, J. Gerritsen, and J.R. Karr. 1996. *Biological Criteria: Technical Guidance for Streams and Small Rivers*. EPA 822-B-96-001. Washington, DC: U.S. Environmental Protection Agency.

Gore, J.A., A. Middleton, D.L. Hughes, U. Rai, and M.P. Brossett. 2006. *A Numerical Index of Health of Wadeable Streams in Georgia Using a Multimetric Index for Benthic Macroinvertebrates*. Atlanta: Georgia Department of Natural Resources.

Gore, J.A., J.R. Olson, D.L. Hughes, and M. Brossett. 2005. *Reference Conditions for Wadeable Streams in Georgia with a Multimetric Index for the Bioassessment and Discrimination of Reference and Impaired Streams*. Atlanta: Georgia Department of Natural Resources.

Ohio EPA. 1999. *Total Maximum Daily Load*. Columbus: Ohio Environmental Protection Agency, Division of Surface Water.

Stribling, J.B., K.L. Pavlik, S.M. Holdsworth, and E.W. Leppo. 2008. Data quality, performance, and uncertainty in taxonomic identification for biological assessment. *Journal of North American Benthological Society* 27(4): 906–919.

Yoder, C.O., and M.T. Barbour. 2009. Critical technical elements of state bioassessment programs: A process to evaluate program rigor and comparability. *Environmental Monitoring Assessment* 150: 31–42.

Appendix A: Selected 1998 Georgia Land Use Values for All Stream Sites

Station ID	Subecoregion	Condition	% Urban	% Agriculture Pasture	% Agriculture Row Crop	% Agriculture Total	% Barren
45a-31	45a	Impaired	8.9	27.8	1.1	28.9	12.4
45a-35	45a	Impaired	16.4	23.5	0.0	23.5	4.9
45a-50	45a	Impaired	55.6	2.0	0.4	2.3	5.5
45a-59	45a	Impaired	78.4	0.5	0.1	0.6	0.0
45a-61	45a	Impaired	59.7	3.5	0.5	4.0	1.7
45a-90	45a	Impaired	5.1	36.3	4.3	40.6	7.4
45a03//	45a	Reference	6.7	3.6	0.0	3.6	7.0
45a-3	45a	Reference	4.9	5.2	0.0	5.3	6.5
45a-89	45a	Reference	0.0	6.6	7.2	13.8	0.5
HH16	45a	Reference	7.2	27.4	2.4	29.8	11.7
HH18	45a	Reference	8.8	16.4	1.2	17.7	11.1
45b-120	45b	Impaired	9.0	52.3	7.4	59.7	3.4
45b-193	45b	Impaired	56.3	6.7	0.2	6.8	2.5
45b-203	45b	Impaired	77.0	0.5	0.0	0.5	1.3
45b-217	45b	Impaired	77.4	0.6	0.0	0.6	3.3
45b-291	45b	Impaired	81.9	0.7	0.0	0.7	2.2
45b-44	45b	Impaired	12.2	15.4	0.5	15.9	8.9
45b-152	45b	Reference	4.0	1.2	0.1	1.2	3.0
45b-156	45b	Reference	4.9	3.0	0.2	3.2	5.5
45b-258	45b	Reference	6.5	11.7	2.3	14.0	5.9
45b-357	45b	Reference	4.8	2.6	0.3	2.8	14.7
HH22	45b	Reference	6.4	10.9	0.8	11.7	4.6
45c-10	45c	Impaired	6.1	31.9	1.8	33.8	7.2
45c-11	45c	Impaired	6.8	26.2	2.0	28.2	5.5
45c-17	45c	Impaired	4.8	20.1	1.7	21.7	9.7
45c-18	45c	Ref. Removed	17.5	14.8	1.3	16.0	10.5
45c-3	45c	Impaired	4.5	30.2	1.5	31.6	13.4
45c-7	45c	Impaired	6.7	25.4	4.2	29.6	8.9
//4	45c	Reference	5.9	12.0	0.7	12.7	14.9

Note: Sites with // in the Station ID are Best Professional Judgment (BPJ) sites from Georgia Environmental Protection Division. Sites with HH in the Station ID are BPJ sites from the U.S. Environmental Protection Agency.

(*Continued*)

Station ID	Subecoregion	Condition	% Urban	% Agriculture Pasture	% Agriculture Row Crop	% Agriculture Total	% Barren
45c-16	45c	Reference	4.0	16.3	1.1	17.4	6.8
45c-19	45c	Reference	3.3	1.1	0.0	1.1	7.3
45c-8	45c	Reference	5.5	11.4	0.9	12.3	13.1
HH24	45c	Reference	5.2	7.7	8.4	16.1	21.1
45d-11	45d	Impaired	5.5	24.2	0.6	24.8	10.0
45d-14	45d	Impaired	6.2	9.1	0.7	9.8	11.5
45d-21	45d	Impaired	4.8	16.1	1.0	17.0	9.7
45d-23	45d	Impaired	14.5	7.5	0.7	8.2	6.6
45d-6	45d	Impaired	4.5	2.7	0.0	2.7	3.3
45d-8	45d	Ref. Removed	5.5	5.6	0.7	6.4	7.8
45d-15	45d	Reference	0.1	0.5	0.7	1.2	6.3
45d-16	45d	Reference	0.1	2.1	1.3	3.4	1.6
45d-4	45d	Reference	8.7	13.5	0.0	13.5	11.0
45d-9	45d	Reference	2.8	3.2	0.2	3.4	5.5
45h-1	45h	Impaired	5.9	10.1	2.0	12.1	5.9
45h-10	45h	Impaired	6.1	21.7	5.8	27.5	4.2
45h-11	45h	Impaired	17.1	7.7	1.5	9.2	4.2
45h-12	45h	Impaired	8.3	10.4	3.0	13.4	12.3
45h-2	45h	Impaired	5.0	10.7	3.9	14.6	6.8
45h-13	45h	Reference	6.7	3.8	0.6	4.4	7.6
45h-16	45h	Reference	7.2	0.0	0.0	0.0	18.7
45h-17	45h	Reference	7.0	0.0	0.0	0.0	3.6
45h-6	45h	Reference	4.8	7.8	0.8	8.6	5.3
45h-9	45h	Reference	5.1	13.3	2.4	15.7	3.1
65c-12	65c	Impaired	6.0	11.2	5.7	16.9	5.5
65c-3	65c	Impaired	48.0	6.2	3.8	10.1	7.2
65c-4	65c	Impaired	59.9	3.3	4.3	7.7	7.0
65c-40	65c	Impaired	3.5	12.3	13.8	26.1	21.5
65c-5	65c	Impaired	35.2	5.1	3.4	8.5	6.1
65c-8	65c	Impaired	10.2	14.1	17.1	31.2	5.6
65c-80	65c	Reference	3.0	0.2	9.3	9.5	17.4
65c-89	65c	Reference	7.3	0.6	0.0	0.7	15.3
HH24	65c	Reference	5.2	7.7	8.4	16.1	21.1
HH25	65c	Reference	4.6	4.4	4.5	8.9	13.5
HH26	65c	Reference	5.6	8.0	13.1	21.2	9.0
65d-1	65d	Impaired	81.9	0.2	0.2	0.4	3.8
65d-20	65d	Impaired	4.6	0.8	3.7	4.5	20.0
65d-21	65d	Impaired	5.1	3.9	23.7	27.7	7.5
65d-32	65d	Impaired	5.4	1.4	18.8	20.1	9.3
65d-39	65d	Impaired	5.1	4.6	23.4	28.0	7.0
65d-14	65d	Reference	5.3	1.1	5.7	6.8	9.9
65d-18	65d	Reference	5.2	0.1	0.4	0.5	15.1

(Continued)

Station ID	Subecoregion	Condition	% Urban	% Agriculture Pasture	% Agriculture Row Crop	% Agriculture Total	% Barren
65d-3	65d	Reference	5.4	0.9	0.0	0.9	16.2
65d-38	65d	Reference	3.4	0.5	3.6	4.1	14.6
65d-4	65d	Reference	4.2	0.1	0.0	0.1	18.1
65g-10	65g	Impaired	6.6	0.0	64.0	64.0	5.8
65g-130	65g	Impaired	5.3	6.9	59.2	66.1	5.3
65g-135	65g	Impaired	14.6	5.3	53.6	58.9	5.0
65g-137	65g	Impaired	6.6	9.0	55.9	64.9	6.1
65g-14	65g	Impaired	8.8	0.0	65.1	65.1	2.2
65g-17	65g	Impaired	9.3	0.0	68.0	68.0	3.5
65g-4	65g	Impaired	5.2	0.0	76.9	76.9	2.3
65g-69	65g	Impaired	29.6	0.0	35.6	35.6	2.9
65g-8	65g	Impaired	7.7	0.0	73.1	73.1	1.4
65g-84	65g	Impaired	6.1	0.7	59.4	60.1	4.9
65g-82	65g	Ref. Removed	5.2	3.9	27.1	31.0	4.1
65g-83	65g	Ref. Removed	6.2	3.3	11.1	14.4	6.8
65g-120	65g	Reference	6.7	7.3	23.4	30.7	8.3
65g-62	65g	Reference	5.4	1.1	36.5	37.6	2.5
HH29	65g	Reference	5.5	5.0	36.2	41.2	4.6
65h-17	65h	Impaired	4.7	0.9	59.0	59.9	14.4
65h-174	65h	Impaired	32.9	5.2	25.2	30.4	1.4
65h-32	65h	Impaired	7.1	3.9	61.5	65.3	6.9
65h-34	65h	Impaired	5.8	2.4	70.1	72.5	5.2
65h-41	65h	Impaired	5.6	0.0	66.9	66.9	2.1
65h-5	65h	Impaired	6.6	0.0	66.6	66.6	5.1
65h-202	65h	Reference	8.3	5.3	21.8	27.1	6.5
65h-203	65h	Reference	8.7	3.0	9.5	12.5	11.0
65h-206	65h	Reference	0.1	2.7	22.1	24.8	5.7
65h-209	65h	Reference	7.2	13.8	32.3	46.1	3.0
65h-212	65h	Reference	1.6	2.8	18.4	21.2	6.5
65k-102	65k	Impaired	5.1	10.3	70.7	81.1	0.7
65k-113	65k	Impaired	25.5	0.0	32.4	32.4	4.0
65k-128	65k	Impaired	5.6	4.0	42.1	46.2	2.3
65k-129	65k	Impaired	5.3	2.7	30.9	33.6	5.1
65k-37	65k	Impaired	7.1	4.9	36.6	41.5	3.1
65k-54	65k	Reference	5.4	2.5	14.8	17.3	1.4
65k-55	65k	Reference	5.6	1.9	25.2	27.1	1.8
65k-56	65k	Reference	4.5	2.5	15.3	17.8	10.6
65k-68	65k	Reference	6.8	0.5	10.5	11.0	7.7
65k-85	65k	Reference	11.8	2.0	10.8	12.8	7.3
65l-160	65l	Impaired	31.1	6.9	26.6	33.6	2.5
65l-184	65l	Impaired	5.3	1.8	28.2	30.0	8.8

(Continued)

Station ID	Subecoregion	Condition	% Urban	% Agriculture Pasture	% Agriculture Row Crop	% Agriculture Total	% Barren
65l-391	65l	Impaired	6.2	1.9	52.4	54.3	7.2
65l-420	65l	Impaired	4.7	5.9	59.0	64.9	3.5
65l-423	65l	Impaired	12.8	7.2	30.8	38.0	5.4
65l-10	65l	Reference	4.3	8.0	13.8	21.8	2.9
65l-342	65l	Reference	6.6	8.3	21.9	30.3	4.8
65l-343	65l	Reference	5.5	4.7	21.6	26.3	8.6
65l-379	65l	Reference	2.4	0.1	2.8	2.9	5.5
65l-381	65l	Reference	7.9	0.5	10.7	11.3	24.8
65o-11	65o	Impaired	5.7	2.2	5.2	7.4	2.3
65o-18	65o	Impaired	0.1	1.0	22.0	22.9	14.2
65o-22	65o	Impaired	0.2	1.5	19.1	20.6	21.9
65o-3	65o	Impaired	27.0	3.3	7.6	11.0	8.9
65o-9	65o	Impaired	2.8	4.8	12.0	16.9	7.2
65o-12	65o	Reference	7.0	1.5	3.3	4.8	0.2
65o-23	65o	Reference	6.3	8.8	32.3	41.1	4.0
65o-24	65o	Reference	5.2	7.8	36.9	44.8	5.5
65o-25	65o	Reference	5.5	2.2	12.9	15.1	2.9
66d-38	66d	Impaired	4.8	0.5	0.0	0.5	0.1
66d-43	66d	Impaired	2.4	6.6	0.0	6.6	3.2
66d-48	66d	Impaired	1.0	2.3	0.2	2.5	0.1
66d-49	66d	Impaired	15.6	0.3	9.0	9.4	2.6
66d-50	66d	Impaired	24.2	1.4	5.8	7.2	3.0
66d-40	66d	Reference	1.9	0.2	0.0	0.2	0.1
66d-41	66d	Reference	2.6	0.2	0.0	0.2	0.2
66d-44	66d	Reference	0.0	0.0	0.0	0.0	0.0
66d-44-2	66d	Reference	0.0	0.0	0.0	0.0	0.0
66d-58	66d	Reference	1.8	0.3	0.0	0.3	0.5
66g-30	66g	Impaired	28.7	12.2	0.0	12.2	3.0
66g-31	66g	Impaired	15.1	9.6	0.0	9.6	4.8
66g-39	66g	Impaired	14.4	21.5	0.0	21.5	10.2
66g-42	66g	Impaired	7.9	3.7	0.0	3.7	6.8
66g-44	66g	Impaired	6.3	25.6	0.0	25.6	6.9
66g-65	66g	Impaired	9.6	6.9	0.0	6.9	4.5
66g-71	66g	Impaired	7.5	8.8	0.0	8.8	12.0
66g-2	66g	Reference	1.5	0.2	0.0	0.2	0.1
66g-2-2	66g	Reference	1.3	0.0	0.0	0.0	0.0
66g-23	66g	Reference	2.5	0.0	0.0	0.0	0.1
66g-5	66g	Reference	3.9	0.2	0.0	0.2	0.0
66g-6	66g	Reference	3.9	0.2	0.0	0.2	0.0
66j-17	66j	Impaired	2.1	4.6	0.0	4.6	0.7
66j-25	66j	Impaired	8.1	18.1	0.0	18.1	0.3
66j-26	66j	Impaired	0.5	3.8	0.9	4.7	2.6
66j-27	66j	Impaired	5.9	10.2	0.0	10.2	0.8

(Continued)

Station ID	Subecoregion	Condition	% Urban	% Agriculture Pasture	% Agriculture Row Crop	% Agriculture Total	% Barren
66j-9	66j	Impaired	7.0	21.8	0.0	21.8	4.6
66j-19	66j	Reference	0.3	2.4	0.5	2.9	0.4
66j-211	66j	Reference	4.2	10.4	0.0	10.4	0.1
66j-23	66j	Reference	0.1	8.9	0.6	9.6	0.0
66j-28	66j	Reference	0.1	3.0	0.4	3.4	0.0
66j-31	66j	Reference	4.1	4.4	0.0	4.4	4.0
67f&i-1	67f&i	Impaired	53.7	7.5	0.0	7.5	3.5
67f&i-11	67f&i	Impaired	21.7	45.7	0.0	45.7	7.0
67f&i-20	67f&i	Impaired	13.8	46.9	0.0	46.9	2.2
67f&i-33	67f&i	Impaired	5.7	30.3	0.0	30.3	3.1
67f&i-5	67f&i	Impaired	39.2	22.8	0.0	22.8	4.5
67f&i-16	67f&i	Reference	5.8	26.3	0.0	26.3	1.6
67f&i-17	67f&i	Reference	5.4	17.4	0.0	17.4	5.1
67f&i-25	67f&i	Reference	2.8	0.1	0.0	0.1	3.7
67f&i-27	67f&i	Reference	7.2	15.2	0.0	15.2	3.0
67f&i-37	67f&i	Reference	5.4	9.2	0.0	9.2	17.8
67g-1	67g	Impaired	0.2	28.5	5.3	33.9	3.1
67g-19	67g	Impaired	22.2	25.3	0.0	25.3	1.8
67g-6	67g	Impaired	6.7	15.1	0.0	15.1	6.2
67g-7	67g	Impaired	5.9	7.6	0.0	7.6	5.1
67g-9	67g	Impaired	7.2	20.1	0.0	20.1	9.4
67g-2	67g	Ref. Removed	9.7	17.6	0.0	17.6	2.3
67g-11	67g	Reference	4.0	20.4	0.0	20.4	0.4
67g-12	67g	Reference	4.2	30.4	0.0	30.4	0.8
67g-13	67g	Reference	3.8	19.5	0.0	19.5	3.6
67g-15	67g	Reference	8.8	12.7	0.0	12.7	2.7
67h-5	67h	Impaired	1.8	0.0	0.0	0.0	4.0
67h-8	67h	Impaired	0.0	1.4	0.0	1.4	0.2
67h-2	67h	Reference	4.2	9.3	0.0	9.3	6.1
67h-3	67h	Reference	4.5	8.0	0.0	8.0	3.1
67h-4	67h	Reference	2.4	0.6	0.0	0.6	1.6
67h-9	67h	Reference	6.4	10.6	0.0	10.6	5.1
68c&d-1	68c&d	Impaired	2.8	2.3	1.6	3.9	0.0
68c&d-10	68c&d	Impaired	4.3	21.9	0.0	21.9	2.0
68c&d-3	68c&d	Impaired	8.4	43.4	0.0	43.4	0.1
68c&d-7	68c&d	Impaired	2.0	14.2	0.0	14.2	3.3
68c&d-8	68c&d	Impaired	0.0	2.4	0.5	2.8	0.0
68c&d-4	68c&d	Reference	9.7	9.6	0.0	9.6	1.5
68c&d-5	68c&d	Reference	8.3	18.3	0.0	18.3	1.4
68c&d-6	68c&d	Reference	7.2	25.2	0.0	25.2	2.6
68c&d-9	68c&d	Reference	3.0	17.0	0.0	17.0	0.6
75e-20	75e	Impaired	7.0	0.3	2.0	2.3	18.3

(Continued)

Station ID	Subecoregion	Condition	% Urban	% Agriculture Pasture	% Agriculture Row Crop	% Agriculture Total	% Barren
75e-3	75e	Impaired	3.9	1.5	24.1	25.6	6.9
75e-36	75e	Impaired	5.6	1.9	32.6	34.5	15.1
75e-46	75e	Impaired	9.1	9.1	23.4	32.5	12.9
75e-54	75e	Impaired	7.9	0.2	0.4	0.6	18.2
75e-23	75e	Reference	5.0	0.7	7.0	7.6	16.0
75e-59	75e	Reference	6.8	0.2	1.7	1.9	15.0
75e-60	75e	Reference	5.0	1.1	3.0	4.1	8.3
75e-69	75e	Reference	9.1	0.0	12.1	12.1	29.7
75e-78	75e	Reference	6.2	1.7	4.5	6.2	8.4
75f-127	75f	Impaired	22.1	2.2	4.4	6.6	27.1
75f-137	75f	Impaired	42.2	1.0	1.6	2.6	26.9
75f-44	75f	Impaired	75.0	0.3	0.3	0.5	5.4
75f-45	75f	Impaired	84.7	0.0	0.0	0.0	3.9
75f-50	75f	Impaired	15.5	4.9	25.8	30.8	12.6
75f-124	75f	Reference	5.5	0.9	0.6	1.5	11.3
75f-126	75f	Reference	9.3	0.0	0.0	0.0	12.2
75f-61	75f	Reference	6.5	0.0	0.0	0.0	4.8
75f-91	75f	Reference	6.0	0.0	0.0	0.1	12.5
75f-95	75f	Reference	7.8	0.0	0.2	0.2	6.9
75h-1	75h	Impaired	8.5	0.0	33.1	33.1	4.4
75h-41	75h	Impaired	8.1	0.9	2.1	3.1	24.2
75h-47	75h	Impaired	4.2	2.1	13.5	15.6	5.9
75h-69	75h	Impaired	11.4	11.6	24.8	36.4	8.5
75h-70	75h	Impaired	7.9	5.6	20.2	25.7	19.3
75h-72	75h	Impaired	12.1	7.1	11.5	18.6	9.7
75h-10	75h	Reference	5.5	0.0	3.3	3.3	16.0
75h-35	75h	Reference	4.6	3.8	18.5	22.3	3.4
75h-45	75h	Reference	9.8	0.4	3.8	4.2	2.1
75h-60	75h	Reference	8.3	1.1	26.0	27.1	9.0
75h-66	75h	Reference	7.7	7.2	16.2	23.4	22.3
75j-13	75j	Impaired	74.6	0.0	0.0	0.0	3.0
75j-2	75j	Impaired	31.9	3.5	3.4	6.9	12.1
75j-24	75j	Impaired	26.1	0.0	0.0	0.0	10.2
75j-3	75j	Impaired	54.4	0.0	1.0	1.0	4.2
75j-4	75j	Impaired	82.0	0.0	0.0	0.0	2.7
75j-29	75j	Ref. Removed	20.5	3.5	1.6	5.1	5.3
75j-10	75j	Reference	5.3	0.0	0.0	0.0	20.7
75j-15	75j	Reference	4.1	0.0	0.0	0.0	0.2
75j-16	75j	Reference	9.9	0.0	0.0	0.0	8.7
75j-25	75j	Reference	11.0	0.2	0.3	0.5	16.5
75j-26	75j	Reference	20.2	1.3	0.7	2.0	11.4
75j-31	75j	Reference	5.4	0.0	0.0	0.0	19.2
75j-37	75j	Reference	5.3	0.0	0.0	0.0	0.3
75j-41	75j	Reference	8.4	0.2	0.8	1.0	0.6
75j-5	75j	Reference	5.7	0.8	0.1	0.9	35.0

Appendix B: Taxonomic References

Brigham, A.R., U. Brigham, and A. Gnilka, eds. 1982. *Aquatic Insects and Oligochaetes of North and South Carolina*. Mahomet, IL: Midwest Aquatic Enterprises.

Burch, J.B. 1982. *Freshwater Snails (Mollusca: Gastropoda) of North America*. Cincinnati, OH: U.S. Environmental Protection Agency, Office of Research and Development, Environmental Monitoring and Support Laboratory.

Daigle, J.J. 1991. *Florida Damselflies (Zygoptera): A Species Key to the Larval Stages* (Technical Series, Vol. 11, No. 1). Tallahassee: State of Florida, Department of Environmental Regulation.

Daigle, J.J. 1992. *Florida Dragonflies (Anisoptera): A Species Key to the Larval Stages* (Technical Series, Vol. 12, No. 1). Tallahassee: State of Florida, Department of Environmental Regulation.

Epler, J.H. 1996. *Identification Manual for the Water Beetles of Florida*. Tallahassee: Florida Department of Environmental Protection.

Epler, J.H. 2001. *Identification Manual for the Larval Chironomidae (Diptera) of North and South Carolina*.

Hobbs, H.H., Jr. 1981. *The Crayfishes of Georgia* (Smithsonian Contribution to Zoology #318). Washington, DC: Smithsonian Institution Press.

Merritt, R.W., and K.W. Cummins, eds. 1996. *An Introduction to the Aquatic Insects of North America*, 3rd ed. Dubuque, IA: Kendall/Hunt.

Pennak, R.W. 1978. *Freshwater Invertebrates of the United States*, 2nd ed. New York: John Wiley & Sons.

Pescador, M.L., A. Rasmussen, and S. Harris. 1995. *Identification Manual for the Caddisfly (Trichoptera) Larvae of Florida*. Tallahassee: State of Florida, Department of Environmental Protection, Division of Water Facilities.

Pescador, M.L., A. Rasmussen, and B. Richard. 2000. *A Guide to the Stoneflies (Plecoptera) of Florida*. Tallahassee: State of Florida, Department of Environmental Protection, Division of Water Resource Management.

Thorp, J.H., and A. Covich, eds. 1991. *Ecology and Classification of North American Freshwater Invertebrates*. San Diego, CA: Academic Press.

Wiggins, G.B. 1977. *Larvae of the North American Caddisfly Genera (Trichoptera)*. Toronto: University of Toronto Press.

Appendix C: List of Stream Sites

Station ID	Salinity	Stream Order	Ecoregion	Subecoregion	Condition
45a-35	Fresh	3	45	45a	Impaired
45a-50	Fresh	3	45	45a	Impaired
45a-59	Fresh	3	45	45a	Impaired
45a-61	Fresh	3	45	45a	Impaired
45a-90	Fresh	4	45	45a	Impaired
45a-3	Fresh	4	45	45a	Reference
45a-89	Fresh	2	45	45a	Reference
45a03//	Fresh	4	45	45a	Reference-BPJ
HH16	Fresh	2	45	45a	Reference-BPJ
HH18	Fresh	3	45	45a	Reference-BPJ
45b-120	Fresh	1	45	45b	Impaired
45b-193	Fresh	2	45	45b	Impaired
45b-203	Fresh	3	45	45b	Impaired
45b-217	Fresh	3	45	45b	Impaired
45b-291	Fresh	3	45	45b	Impaired
45b-44	Fresh	2	45	45b	Impaired
45b-152	Fresh	3	45	45b	Reference
45b-156	Fresh	2	45	45b	Reference
45b-258	Fresh	4	45	45b	Reference
45b-357	Fresh	3	45	45b	Reference
HH22	Fresh	3	45	45b	Reference-BPJ
45c-10	Fresh	2	45	45c	Impaired
45c-11	Fresh	4	45	45c	Impaired
45c-17	Fresh	4	45	45c	Impaired
45c-3	Fresh	3	45	45c	Impaired
45c-7	Fresh	2	45	45c	Impaired
45c-16	Fresh	3	45	45c	Reference
45c-19	Fresh	3	45	45c	Reference
45c-8	Fresh	4	45	45c	Reference
//4	Fresh	2	45	45c	Reference-BPJ
45d-11	Fresh	1	45	45d	Impaired
45d-14	Fresh	4	45	45d	Impaired
45d-21	Fresh	3	45	45d	Impaired
45d-23	Fresh	2	45	45d	Impaired
45d-6	Fresh	4	45	45d	Impaired
45d-15	Fresh	4	45	45d	Reference

(Continued)

Station ID	Salinity	Stream Order	Ecoregion	Subecoregion	Condition
45d-16	Fresh	2	45	45d	Reference
45d-4	Fresh	2	45	45d	Reference
45d-9	Fresh	3	45	45d	Reference
45h-1	Fresh	3	45	45h	Impaired
45h-10	Fresh	3	45	45h	Impaired
45h-11	Fresh	3	45	45h	Impaired
45h-12	Fresh	3	45	45h	Impaired
45h-2	Fresh	2	45	45h	Impaired
45h-13	Fresh	3	45	45h	Reference
45h-16	Fresh	2	45	45h	Reference
45h-17	Fresh	3	45	45h	Reference
45h-6	Fresh	3	45	45h	Reference
45h-9	Fresh	3	45	45h	Reference
65c-12	Fresh	2	65	65c	Impaired
65c-3	Fresh	3	65	65c	Impaired
65c-4	Fresh	4	65	65c	Impaired
65c-40	Fresh	3	65	65c	Impaired
65c-5	Fresh	3	65	65c	Impaired
65c-8	Fresh	3	65	65c	Impaired
65c-88	Fresh	3	65	65c	Impaired
65c-80	Fresh	3	65	65c	Reference
65c-89	Fresh	4	65	65c	Reference
HH24	Fresh	3	65	65c	Reference-BPJ
HH25	Fresh	4	65	65c	Reference-BPJ
HH26	Fresh	3	65	65c	Reference-BPJ
65d-1	Fresh	3	65	65d	Impaired
65d-20	Fresh	1	65	65d	Impaired
65d-21	Fresh	3	65	65d	Impaired
65d-32	Fresh	2	65	65d	Impaired
65d-39	Fresh	3	65	65d	Impaired
65d-14	Fresh	4	65	65d	Reference
65d-18	Fresh	4	65	65d	Reference
65d-3	Fresh	3	65	65d	Reference
65d-38	Fresh	3	65	65d	Reference
65d-4	Fresh	3	65	65d	Reference
65g-10	Fresh	2	65	65g	Impaired
65g-130	Fresh	2	65	65g	Impaired
65g-135	Fresh	3	65	65g	Impaired
65g-137	Fresh	2	65	65g	Impaired
65g-14	Fresh	2	65	65g	Impaired
65g-17	Fresh	3	65	65g	Impaired
65g-4	Fresh	2	65	65g	Impaired
65g-69	Fresh	3	65	65g	Impaired
65g-8	Fresh	2	65	65g	Impaired
65g-84	Fresh	1	65	65g	Impaired

(Continued)

Station ID	Salinity	Stream Order	Ecoregion	Subecoregion	Condition
65g-120	Fresh	3	65	65g	Reference
65g-62	Fresh	4	65	65g	Reference
HH29	Fresh	4	65	65g	Reference-BPJ
65h-17	Fresh	2	65	65h	Impaired
65h-174	Fresh	2	65	65h	Impaired
65h-32	Fresh	3	65	65h	Impaired
65h-34	Fresh	1	65	65h	Impaired
65h-41	Fresh	3	65	65h	Impaired
65h-202	Fresh	3	65	65h	Reference
65h-203	Fresh	3	65	65h	Reference
65h-206	Fresh	2	65	65h	Reference
65h-209	Fresh	3	65	65h	Reference
65h-212	Fresh	3	65	65h	Reference
65k-102	Fresh	3	65	65k	Impaired
65k-113	Fresh	2	65	65k	Impaired
65k-128	Fresh	2	65	65k	Impaired
65k-129	Fresh	3	65	65k	Impaired
65k-37	Fresh	4	65	65k	Impaired
65k-54	Fresh	3	65	65k	Reference
65k-55	Fresh	3	65	65k	Reference
65k-56	Fresh	4	65	65k	Reference
65k-68	Fresh	3	65	65k	Reference
65k-85	Fresh	4	65	65k	Reference
65l-160	Fresh	2	65	65l	Impaired
65l-184	Fresh	2	65	65l	Impaired
65l-391	Fresh	3	65	65l	Impaired
65l-420	Fresh	2	65	65l	Impaired
65l-423	Fresh	3	65	65l	Impaired
65l-10	Fresh	3	65	65l	Reference
65l-342	Fresh	2	65	65l	Reference
65l-343	Fresh	2	65	65l	Reference
65l-379	Fresh	2	65	65l	Reference
65l-381	Fresh	3	65	65l	Reference
65o-11	Fresh	1	65	65o	Impaired
65o-18	Fresh	3	65	65o	Impaired
65o-22	Fresh	2	65	65o	Impaired
65o-3	Fresh	3	65	65o	Impaired
65o-9	Fresh	1	65	65o	Impaired
65o-12	Fresh	2	65	65o	Reference
65o-23	Fresh	2	65	65o	Reference
65o-24	Fresh	2	65	65o	Reference
65o-25	Fresh	2	65	65o	Reference
66d-38	Fresh	3	66	66d	Impaired
66d-43	Fresh	4	66	66d	Impaired
66d-48	Fresh	2	66	66d	Impaired

(Continued)

Station ID	Salinity	Stream Order	Ecoregion	Subecoregion	Condition
66d-49	Fresh	2	66	66d	Impaired
66d-50	Fresh	3	66	66d	Impaired
66d-40	Fresh	4	66	66d	Reference
66d-41	Fresh	4	66	66d	Reference
66d-44	Fresh	4	66	66d	Reference
66d-44-2	Fresh	2	66	66d	Reference
66d-58	Fresh	4	66	66d	Reference
66g-30	Fresh	3	66	66g	Impaired
66g-31	Fresh	4	66	66g	Impaired
66g-39	Fresh	3	66	66g	Impaired
66g-42	Fresh	3	66	66g	Impaired
66g-44	Fresh	2	66	66g	Impaired
66g-65	Fresh	3	66	66g	Impaired
66g-71	Fresh	4	66	66g	Impaired
66g-2	Fresh	3	66	66g	Reference
66g-2-2	Fresh	3	66	66g	Reference
66g-23	Fresh	3	66	66g	Reference
66g-5	Fresh	4	66	66g	Reference
66g-6	Fresh	4	66	66g	Reference
66j-17	Fresh	3	66	66j	Impaired
66j-25	Fresh	3	66	66j	Impaired
66j-26	Fresh	3	66	66j	Impaired
66j-27	Fresh	3	66	66j	Impaired
66j-9	Fresh	3	66	66j	Impaired
66j-19	Fresh	3	66	66j	Reference
66j-211	Fresh	3	66	66j	Reference
66j-23	Fresh	3	66	66j	Reference
66j-28	Fresh	3	66	66j	Reference
66j-31	Fresh	3	66	66j	Reference
67f&i-1	Fresh	4	67	67f&i	Impaired
67f&i-11	Fresh	2	67	67f&i	Impaired
67f&i-20	Fresh	4	67	67f&i	Impaired
67f&i-33	Fresh	3	67	67f&i	Impaired
67f&i-5	Fresh	4	67	67f&i	Impaired
67f&i-16	Fresh	3	67	67f&i	Reference
67f&i-17	Fresh	5	67	67f&i	Reference
67f&i-25	Fresh	2	67	67f&i	Reference
67f&i-27	Fresh	4	67	67f&i	Reference
67f&i-37	Fresh	4	67	67f&i	Reference
67g-1	Fresh	3	67	67g	Impaired
67g-19	Fresh	4	67	67g	Impaired
67g-6	Fresh	4	67	67g	Impaired
67g-7	Fresh	3	67	67g	Impaired
67g-9	Fresh	4	67	67g	Impaired
67g-11	Fresh	3	67	67g	Reference

(*Continued*)

Station ID	Salinity	Stream Order	Ecoregion	Subecoregion	Condition
67g-12	Fresh	4	67	67g	Reference
67g-13	Fresh	3	67	67g	Reference
67g-15	Fresh	3	67	67g	Reference
67h-5	Fresh	3	67	67h	Impaired
67h-8	Fresh	3	67	67h	Impaired
67h-2	Fresh	4	67	67h	Reference
67h-3	Fresh	4	67	67h	Reference
67h-4	Fresh	2	67	67h	Reference
67h-9	Fresh	2	67	67h	Reference
68c&d-1	Fresh	4	68	68c&d	Impaired
68c&d-10	Fresh	4	68	68c&d	Impaired
68c&d-3	Fresh	4	68	68c&d	Impaired
68c&d-7	Fresh	3	68	68c&d	Impaired
68c&d-8	Fresh	1	68	68c&d	Impaired
68c&d-4	Fresh	4	68	68c&d	Reference
68c&d-5	Fresh	4	68	68c&d	Reference
68c&d-6	Fresh	4	68	68c&d	Reference
68c&d-9	Fresh	4	68	68c&d	Reference
75e-54	Fresh	2	75	75e	Impaired
75e-46	Fresh	1	75	75e	Impaired
75e-36	Fresh	1	75	75e	Impaired
75e-20	Fresh	2	75	75e	Impaired
75e-36	Fresh	2	75	75e	Impaired
75e-78	Fresh	3	75	75e	Reference
75e-69	Fresh	1	75	75e	Reference
75e-60	Fresh	3	75	75e	Reference
75e-59	Fresh	2	75	75e	Reference
75e-23	Fresh	2	75	75e	Reference
75f-127	Fresh	1	75	75f	Impaired
75f-137	Fresh	3	75	75f	Impaired
75f-44	Fresh	2	75	75f	Impaired
75f-45	Fresh	2	75	75f	Impaired
75f-50	Fresh	2	75	75f	Impaired
75f-124	Saline	2	75	75f	Reference
75f-126	Fresh	2	75	75f	Reference
75f-61	Fresh	3	75	75f	Reference
75f-91	Fresh	3	75	75f	Reference
75f-95	Fresh	2	75	75f	Reference
75h-1	Fresh	1	75	75h	Impaired
75h-41	Fresh	2	75	75h	Impaired
75h-47	Fresh	2	75	75h	Impaired
75h-69	Fresh	2	75	75h	Impaired
75h-70	Fresh	1	75	75h	Impaired
75h-72	Fresh	3	75	75h	Impaired
75h-10	Fresh	3	75	75h	Reference

(Continued)

Station ID	Salinity	Stream Order	Ecoregion	Subecoregion	Condition
75h-35	Fresh	3	75	75h	Reference
75h-45	Fresh	2	75	75h	Reference
75h-60	Fresh	1	75	75h	Reference
75h-66	Fresh	2	75	75h	Reference
75j-13	Fresh	2	75	75j	Impaired
75j-2	Fresh	1	75	75j	Impaired
75j-24	Saline	1	75	75j	Impaired
75j-3	Saline	1	75	75j	Impaired
75j-4	Fresh	2	75	75j	Impaired
75j-10	Fresh	1	75	75j	Reference
75j-15	Fresh	1	75	75j	Reference
75j-16	Saline	1	75	75j	Reference
75j-25	Fresh	2	75	75j	Reference
75j-26	Fresh	1	75	75j	Reference
75j-31	Saline	1	75	75j	Reference
75j-37	Saline	1	75	75j	Reference
75j-41	Saline	2	75	75j	Reference
75j-5	Saline	2	75	75j	Reference

Note: BPJ = Best personal judgment.

Appendix D: Discrimination Efficiencies for Metrics Considered for Index Development

Metric	45	45a	45b	45c	45d	45h	65	65c	65d	65g	65h	65k	65l	65o
TotalTax	0.00	0.00	0.33	0.20	0.00	0.00	0.43	0.14	0.20	0.90	0.60	0.00	0.40	0.40
EPTTax	0.50	0.60	0.67	0.20	0.40	0.40	0.45	0.43	0.20	1.00	0.40	0.00	0.60	0.40
EphemTax	0.35	0.20	0.67	0.20	0.40	0.20	0.43	0.29	0.40	1.00	0.80	0.00	0.00	0.60
PlecoTax	0.31	0.60	0.50	0.00	0.20	1.00	0.00	0.71	0.60	1.00	0.00	0.00	0.00	0.00
TrichTax	0.23	0.40	0.67	0.00	0.00	0.00	0.36	0.43	0.60	1.00	0.20	0.00	0.80	0.20
ColeoTax	0.46	0.40	1.00	0.20	0.80	0.20	0.36	0.14	0.40	0.70	0.80	0.00	0.40	0.20
DipTax	0.08	0.00	0.00	0.00	0.00	0.20	0.24	0.14	0.00	0.70	0.60	0.00	0.80	0.60
ChiroTax	0.12	0.20	0.00	0.00	0.00	0.20	0.21	0.14	0.40	0.60	0.60	0.00	0.60	0.80
TanytTax	0.04	0.00	0.00	0.60	0.00	0.00	0.17	0.14	0.00	0.90	0.80	0.00	0.00	0.80
Evenness	0.19	0.00	0.50	0.60	0.00	0.20	0.45	0.29	0.40	0.50	0.60	0.20	0.60	0.60
Margalef	0.04	0.00	0.33	0.20	0.00	0.20	0.48	0.29	0.20	0.90	0.60	0.00	0.40	0.40
Shan_base_e	0.15	0.00	0.50	0.60	0.00	0.20	0.36	0.29	0.40	0.60	0.60	0.20	0.20	0.40
Simpsons	0.27	0.20	0.50	0.60	0.00	0.20	0.31	0.43	0.40	0.50	0.40	0.20	0.60	0.60
EPTPct	0.54	0.60	0.67	0.40	0.00	0.80	0.45	0.29	0.40	1.00	0.40	0.00	0.60	0.60
EphemPct	0.35	0.40	0.67	0.00	0.00	0.80	0.50	0.29	0.60	0.80	0.80	0.00	0.00	0.80
AmphPct	0.00	0.00	0.00	0.20	0.00	0.00	0.00	0.00	0.00	0.70	0.00	0.20	0.20	0.00
ChiroPct	0.69	1.00	1.00	0.40	0.40	0.60	0.17	0.00	0.60	0.40	0.20	0.40	0.20	0.00
ColeoPct	0.38	0.40	0.83	0.20	0.20	0.20	0.48	0.43	0.40	0.60	0.40	0.40	0.60	0.80
DipPct	0.65	1.00	0.83	0.40	0.00	0.80	0.00	0.14	0.40	0.50	0.20	0.40	0.20	0.40
GastrPct	0.00	0.00	0.00	0.00	0.80	0.00	0.00	0.00	0.00	0.70	0.00	0.80	0.00	0.60
IsoPct	0.00	0.00	0.00	0.00	0.00	0.00	0.31	0.00	0.20	0.30	0.80	0.20	0.20	0.20
NonInPct	0.31	0.40	0.33	0.20	0.00	0.40	0.38	0.29	0.40	0.60	0.60	0.20	0.20	1.00
OdonPct	0.54	0.20	0.33	0.60	0.80	0.60	0.21	0.43	0.40	0.00	0.20	0.40	0.40	0.20
PlecoPct	0.62	0.60	0.67	0.80	0.20	0.80	0.00	0.57	0.40	1.00	0.00	0.00	0.00	0.00
TanytPct	0.19	0.20	0.17	0.40	0.00	0.20	0.14	0.14	0.40	0.90	1.00	0.00	0.00	0.60
OligoPct	0.31	0.40	0.67	0.20	0.20	0.00	0.57	0.29	0.60	1.00	0.80	0.00	0.20	0.80
TrichPct	0.42	0.60	0.67	0.00	0.20	0.20	0.43	0.57	0.60	1.00	0.20	0.00	0.80	0.20
%Orth/TC	0.27	0.00	0.33	0.20	0.60	0.60	0.19	0.14	0.40	0.60	0.00	0.40	0.40	0.00
%Tpod/TC	0.54	0.80	0.17	0.60	1.00	0.80	0.57	0.14	0.20	0.10	0.20	0.60	0.20	0.20
Hyd2TriPct	0.50	0.60	0.50	0.40	0.60	0.40	0.24	0.29	0.80	0.00	0.20	0.60	0.00	0.20
Hyd2EPTPct	0.35	0.60	0.50	0.40	0.60	0.40	0.26	0.14	0.60	0.00	0.20	0.60	0.00	0.20
Tnyt2ChiPct	0.23	0.40	0.33	0.40	0.00	0.20	0.29	0.14	0.40	0.90	1.00	0.00	0.00	0.60
Baet2EphPct	0.00	0.00	0.00	0.00	0.00	0.00	0.00	0.00	0.00	0.00	0.00	0.00	0.00	0.00

(Continued)

Metric	45	45a	45b	45c	45d	45h	65	65c	65d	65g	65h	65k	65l	65o
CrCh2ChiPct	0.42	1.00	0.50	0.40	0.20	0.20	0.33	0.43	0.40	0.50	0.20	0.20	0.40	0.00
TolerTax	0.42	1.00	0.17	0.00	1.00	0.20	0.40	0.71	0.00	0.50	0.40	0.20	0.80	0.00
TolerPct	0.50	1.00	0.50	0.60	1.00	0.40	0.40	0.43	0.20	0.90	0.80	0.00	0.60	0.20
IntolTax	0.42	0.60	0.67	0.20	0.60	0.60	0.55	0.71	0.60	1.00	0.80	0.20	0.00	0.20
IntolPct	0.58	0.60	0.67	0.80	0.80	0.80	0.57	0.43	0.40	1.00	0.40	0.40	0.00	0.20
Dom01Pct	0.19	0.20	0.50	0.60	0.00	0.00	0.33	0.29	0.20	0.30	0.40	0.20	0.20	0.60
Dom01Ind	0.23	0.20	0.50	0.60	0.00	0.00	0.38	0.29	0.20	0.30	0.40	0.20	0.20	0.60
BeckBI	0.46	0.60	0.67	0.20	0.40	0.40	0.57	0.71	0.20	1.00	0.40	0.00	0.00	0.40
HBI	0.50	0.60	0.67	0.80	1.00	1.00	0.45	0.43	0.20	1.00	0.20	0.00	0.60	0.20
NCBI	0.50	1.00	0.50	0.40	1.00	1.00	0.45	0.29	0.40	0.80	0.80	0.20	0.40	0.60
ScrapPct	0.27	0.80	0.50	0.00	0.40	0.60	0.40	0.86	0.40	0.90	0.60	0.40	0.20	0.40
ScrapTax	0.08	0.40	1.00	0.00	0.20	0.20	0.24	0.43	0.20	0.90	0.00	0.60	0.00	0.80
CllctPct	0.08	0.00	0.17	0.00	0.20	0.20	0.10	0.14	0.40	0.00	0.20	0.80	0.20	0.00
CllctTax	0.04	0.00	0.50	0.20	0.00	0.00	0.14	0.00	0.20	0.40	0.00	0.00	0.20	0.40
PredPct	0.08	0.00	0.33	0.20	0.00	0.00	0.48	0.57	0.60	0.70	0.60	0.20	0.20	0.40
PredTax	0.08	0.20	0.33	0.20	0.00	0.00	0.43	0.29	0.40	0.60	0.60	0.00	0.60	0.20
ShredPct	0.35	0.60	0.67	1.00	0.20	0.00	0.29	0.14	0.20	0.30	0.20	0.60	0.00	0.20
ShredTax	0.23	0.60	0.67	0.40	0.40	0.40	0.29	0.29	0.40	0.80	0.20	0.40	0.60	0.20
FiltrPct	0.12	0.20	0.00	0.20	0.60	0.00	0.24	0.43	0.80	0.40	0.20	0.60	0.80	0.40
FiltrTax	0.12	0.00	0.00	0.00	0.00	0.00	0.26	0.29	0.00	0.80	0.80	0.00	0.00	0.60
ClngrTax	0.31	0.80	0.67	0.20	0.00	0.00	0.45	0.43	0.20	1.00	0.80	0.00	0.60	0.40
ClngrPct	0.38	0.80	0.33	0.20	0.00	0.60	0.48	0.43	0.20	0.80	0.80	0.00	0.40	0.40
BrrwrTax	0.00	0.40	0.00	0.20	0.00	0.00	0.40	0.00	0.00	0.40	0.60	0.40	0.20	0.60
ClmbrTax	0.08	0.00	0.33	0.00	0.00	0.40	0.19	0.29	0.20	0.00	0.60	0.00	0.20	0.00
SprwlTax	0.04	0.00	0.33	0.20	0.00	0.20	0.17	0.14	0.40	0.60	0.20	0.00	0.20	0.80
SwmmrTax	0.19	0.20	0.83	0.40	0.00	0.00	0.26	0.00	0.40	0.30	0.20	0.00	0.00	0.40

Metric	66	66d	66g	66j	67	67f&i	67g	67h	68c&d	75	75e	75f	75h	75j
TotalTax	0.53	0.60	0.43	0.60	0.50	0.80	0.40	0.00	0.00	0.15	0.00	0.20	0.33	0.10
EPTTax	0.65	0.80	0.86	0.40	0.75	1.00	0.40	0.00	0.20	0.00	0.00	0.00	0.00	0.00
EphemTax	0.35	0.40	0.71	0.20	0.58	1.00	0.20	1.00	0.60	0.00	0.00	0.00	0.00	0.00
PlecoTax	0.71	0.60	0.43	0.80	0.83	1.00	0.80	0.50	0.80	0.00	0.00	0.00	0.00	0.00
TrichTax	0.53	0.40	0.71	0.40	0.08	0.40	0.00	0.00	0.20	0.00	0.00	0.00	0.00	0.00
ColeoTax	0.18	0.20	0.57	0.20	0.25	0.40	0.20	0.50	0.20	0.27	0.00	0.00	0.17	0.00
DipTax	0.29	0.80	0.00	1.00	0.42	0.40	0.60	0.00	0.00	0.27	0.00	0.40	0.00	0.20
ChiroTax	0.18	0.40	0.00	0.60	0.50	0.80	0.60	0.00	0.00	0.27	0.00	0.40	0.33	0.30
TanytTax	0.00	0.00	0.00	0.60	0.17	0.40	0.40	0.00	0.20	0.00	0.00	0.00	0.00	0.00
Evenness	0.53	0.20	0.43	0.80	0.17	0.40	0.20	0.00	0.00	0.15	0.00	0.00	0.17	0.30
Margalef	0.41	0.60	0.43	0.80	0.50	0.80	0.40	0.00	0.00	0.12	0.00	0.00	0.33	0.10
Shan_base_e	0.53	0.20	0.43	0.80	0.17	0.60	0.20	0.00	0.00	0.15	0.00	0.00	0.17	0.30
Simpsons	0.65	0.20	0.71	0.80	0.17	0.40	0.00	0.00	0.00	0.19	0.00	0.20	0.17	0.30
EPTPct	0.35	0.40	0.43	0.40	0.58	1.00	0.20	0.00	0.40	0.00	0.40	0.00	0.00	0.00
EphemPct	0.06	0.00	0.43	0.20	0.58	1.00	0.40	0.50	0.40	0.00	0.00	0.00	0.00	0.00
AmphPct	0.00	0.00	0.00	0.00	0.00	0.00	0.60	0.00	0.00	0.31	0.60	0.60	0.00	0.30
ChiroPct	0.35	0.60	0.86	0.20	0.50	0.80	0.20	1.00	0.20	0.00	0.00	0.00	0.00	0.10
ColeoPct	0.12	0.20	0.57	0.00	0.17	0.60	0.00	0.50	0.20	0.27	0.20	0.00	0.00	0.00
DipPct	0.29	0.40	0.86	0.40	0.25	0.80	0.00	0.50	0.00	0.04	0.00	0.00	0.00	0.20
GastrPct	0.00	0.00	0.00	0.00	0.58	0.00	0.60	1.00	0.00	0.00	0.00	0.00	0.00	0.00

(Continued)

Metric	45	45a	45b	45c	45d	45h	65	65c	65d	65g	65h	65k	65l	65o
IsoPct	0.00	0.00	0.00	0.00	0.67	0.60	0.60	1.00	0.20	0.27	0.60	0.20	0.50	0.10
NonInPct	0.18	0.20	0.29	0.60	0.42	0.80	0.20	0.00	0.00	0.73	0.80	0.40	1.00	0.40
OdonPct	0.41	1.00	0.00	0.40	0.58	0.60	0.60	0.50	0.60	0.46	0.80	0.80	0.17	0.30
PlecoPct	0.47	0.60	0.57	0.20	0.75	1.00	0.80	0.00	0.60	0.00	0.00	0.00	0.00	0.00
TanytPct	0.00	0.00	0.00	0.80	0.25	0.60	0.60	0.00	0.00	0.00	0.00	0.00	0.00	0.00
OligoPct	0.29	0.20	0.43	0.60	0.50	1.00	0.20	0.00	0.20	0.69	0.80	0.80	0.83	0.50
TrichPct	0.53	0.40	0.71	0.40	0.17	1.00	0.20	0.00	0.20	0.00	0.00	0.00	0.00	0.00
%Orth/TC	0.24	0.40	0.43	0.20	0.58	0.20	1.00	1.00	0.20	0.35	0.00	0.00	0.33	0.00
%Tpod/TC	0.41	0.20	0.86	0.00	0.58	0.60	0.60	0.00	0.80	0.50	0.60	1.00	0.17	0.40
Hyd2TriPct	0.35	0.40	0.29	0.40	0.58	0.60	0.80	0.50	0.60	0.08	0.20	0.00	0.00	0.10
Hyd2EPTPct	0.29	0.40	0.43	0.20	0.58	0.60	0.60	0.50	0.60	0.08	0.20	0.00	0.00	0.10
Tnyt2ChiPct	0.12	0.00	0.29	0.60	0.50	0.60	0.40	0.50	0.20	0.00	0.00	0.00	0.00	0.00
Baet2EphPct	0.00	0.40	0.00	0.00	0.00	0.00	0.80	1.00	0.80	0.00	0.00	0.00	0.00	0.00
CrCh2ChiPct	0.24	0.20	0.29	0.20	0.25	0.60	0.00	0.00	0.40	0.19	0.60	0.00	0.00	0.30
TolerTax	0.18	0.60	1.00	0.00	0.42	0.60	0.40	0.50	0.80	0.23	0.60	0.80	0.00	0.10
TolerPct	0.24	0.20	0.43	0.00	0.08	0.80	0.20	1.00	0.80	0.58	0.40	0.60	0.83	0.50
IntolTax	0.71	0.60	0.71	0.60	0.75	1.00	0.80	0.50	0.80	0.00	0.00	0.00	0.00	0.00
IntolPct	0.59	0.60	0.71	0.60	0.58	0.80	0.40	1.00	1.00	0.00	0.00	0.00	0.00	0.00
Dom01Pct	0.59	0.60	0.71	0.80	0.17	0.40	0.00	0.00	0.00	0.19	0.00	0.20	0.17	0.30
Dom01Ind	0.59	0.40	0.71	0.80	0.08	0.20	0.00	0.00	0.00	0.31	0.40	0.20	0.17	0.30
BeckBI	0.82	0.80	0.71	0.80	0.83	1.00	0.80	0.50	0.60	0.00	0.00	0.00	0.17	0.00
HBI	0.59	0.40	0.71	0.60	0.75	0.80	0.40	1.00	0.80	0.42	0.20	0.60	0.67	0.50
NCBI	0.71	0.60	0.71	0.40	0.75	0.80	0.40	0.50	1.00	0.15	0.20	0.20	0.17	0.70
ScrapPct	0.35	0.40	0.71	0.20	0.33	0.20	0.20	1.00	0.60	0.00	0.00	0.00	0.00	0.00
ScrapTax	0.29	0.00	0.86	0.00	0.58	0.80	0.00	1.00	0.80	0.00	0.00	0.00	0.00	0.00
CllctPct	0.35	0.20	0.29	0.40	0.00	0.20	0.20	0.00	0.00	0.12	0.40	0.00	0.17	0.20
CllctTax	0.24	0.20	0.14	0.60	0.42	0.80	1.00	0.00	0.00	0.23	0.00	0.40	0.50	0.30
PredPct	0.59	0.60	0.71	0.40	0.25	0.20	0.20	0.00	0.20	0.19	0.00	0.00	0.17	0.40
PredTax	0.76	0.60	0.57	0.80	0.17	0.20	0.20	0.00	0.20	0.27	0.00	0.00	0.33	0.50
ShredPct	0.59	0.80	0.14	0.40	0.33	0.60	0.40	0.00	0.40	0.00	0.00	0.00	0.83	0.60
ShredTax	0.76	0.60	0.29	0.80	0.42	0.40	0.80	0.50	0.20	0.00	0.00	0.00	0.67	0.50
FiltrPct	0.59	0.40	0.57	0.80	0.08	0.00	0.00	0.50	0.20	0.35	0.60	0.80	0.50	0.10
FiltrTax	0.12	0.00	0.00	0.40	0.17	0.40	0.20	0.00	0.20	0.00	0.00	0.00	0.00	0.20
ClngrTax	0.53	0.60	0.57	0.20	0.67	1.00	0.20	0.00	0.00	0.00	0.20	0.00	0.00	0.00
ClngrPct	0.35	0.20	0.71	0.20	0.75	1.00	0.40	0.00	0.60	0.00	0.40	0.00	0.00	0.00
BrrwrTax	0.53	0.40	0.57	0.20	0.25	0.00	0.40	0.00	0.20	0.27	0.20	0.40	0.33	0.20
ClmbrTax	0.00	0.00	0.00	0.00	0.00	0.00	0.00	0.00	0.00	0.00	0.00	0.00	0.00	0.00
SprwlTax	0.24	0.20	0.29	0.60	0.42	0.40	0.60	0.00	0.20	0.27	0.20	0.00	0.83	0.30
SwmmrTax	0.18	0.00	0.29	0.00	0.00	0.00	0.20	1.00	0.00	0.00	0.20	0.00	0.33	0.00

Appendix E: Examples of Reference Stream Criteria from the State of Georgia

CONTENTS

ECOREGION 45—PIEDMONT: CHARACTERISTIC REFERENCE STREAM CONDITION

Considered the nonmountainous portion of the old Appalachians Highland by physiographers, the northeast–southwest trending Piedmont ecoregion comprises a transitional area between the mostly mountainous ecoregions of the Appalachians to the northwest and the relatively flat coastal plain to the southeast. It is a complex mosaic of Precambrian and Paleozoic metamorphic and igneous rocks, with moderately dissected irregular plains and some hills. The soils tend to be finer textured than in coastal plain regions (Ecoregions 63 and 65). Once largely cultivated, much of this region has reverted to successional pine and hardwood woodlands, with an increasing conversion to an urban and suburban land cover.

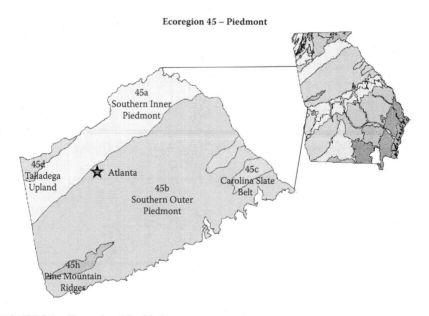

Ecoregion 45 – Piedmont

FIGURE E.1 Ecoregion 45—Piedmont.

TABLE E.1
Characteristic Reference Stream Land Use, Habitat, and Chemistry Data for Ecoregion 45—Piedmont

	Parameter	Mean	Median	Range
Catchment	% Natural	82.7	84.5	40.3–95.6
Land Use	% Agriculture	9.0	7.0	0–29.8
	% Silviculture	8.4	7.0	3.0–18.7
	% Urban	5.7	5.7	2.8–8.8
Habitat	Total Habitat Score (200)	155.6	160.0	128–184
	Epifaunal Substrate (20)	15.7	16.0	8–19
	Embeddedness or Pool Substrate Characterization (20)	14.4	15.0	5–18
	Velocity/Depth Regime or Pool Variability (20)	15.1	16.0	8–19
	Sediment Deposition (20)	13.5	14.0	6–19
	Channel Flow Status (20)	14.7	15.0	7–18
	Channel Alteration (20)	17.5	18.0	15–19
	Frequency of Riffles or Channel Sinuosity (20)	15.4	16.0	9–20
	Bank Stability (L) (10)	7.5	8.0	5–10
	Bank Stability (R) (10)	7.8	8.0	5–10
	Vegetative Protection (L) (10)	8.1	8.0	4–9
	Vegetative Protection (R) (10)	8.1	9.0	4–9
	Riparian Vegetative Width (L) (10)	9.0	9.0	7–10
	Riparian Vegetative Width (R) (10)	8.7	9.0	5–10
In-Stream	% Silt/Clay	3.7	1.0	0–42.0
Habitat	% Sand	36.2	26.2	2.8–89.0
(substrate)	% Gravel	34.9	35.0	2.0–78.0
	% Cobble	18.3	14.1	0–50.0
	% Boulder	4.2	2.0	0–33.0
	% Bedrock	2.7	0.0	0–20.6
Chemistry	Specific Conductivity (mS/cm)	0.113	0.052	0.03–1.21
(*in situ*)	Dissolved Oxygen (mg/l)	8.9	8.9	2.31–13.77
	pH (SU)	6.9	6.9	6.5–7.4
	Turbidity (NTU)	7.1	3.9	0–30.6
Chemistry	Alkalinity (mg/l as $CaCO_3$)	27.9	16.9	6.7–88.2
(laboratory)	Total Hardness (mg/l as $CaCO_3$)	26.1	20.1	6.7–87.2
	Ammonia (mg/l as N)	0.10	0.05	BD–0.97
	Nitrate–Nitrite (mg/l as N)	0.03	0.02	BD–0.08
	Total Phosphorous (mg/l as P)	0.31	0.17	BD–1.17
	Copper (mg/l)	0.007	0.007	BD–0.009
	Iron (mg/l)	2.33	1.12	BD–9.79
	Manganese (mg/l)	0.39	0.35	BD–0.77
	Zinc (mg/l)	0.03	0.02	BD–0.07

Note: BD = Below detection.

DISCRIMINATING INVERTEBRATE METRICS FOR THE CHARACTERISTIC REFERENCE STREAM

Approximately 70 invertebrate metrics were evaluated for Ecoregion 45 based upon data collected from candidate reference streams. Table E.2 includes raw data values for metrics that were judged by the strength of their discrimination between reference and impaired sites, and by graphical analysis, to be candidates for inclusion in final indices.

TABLE E.2
Discriminating Invertebrate Metrics for the Characteristic Reference Stream for Ecoregion 45—Piedmont

Metric Category	Metric	Mean	Median	Range
Richness	Ephemeroptera, Plecoptera, Trichoptera Taxa (EPT Taxa)	11.6	11.0	2–20
Composition	% Chironomidae	31.8	33.2	3.0–59.6
	% EPT	39.2	34.6	0.8–74.1
	% Diptera	38.8	41.3	7.6–79.7
	% Odonata	1.7	1.4	0–7.4
	% Plecoptera	13.2	6.9	0.4–74.1
Tolerance/	Hilsenhoff's Biotic Index (HBI)	5.1	5.1	2.1–7.4
Intolerance	% Intolerant Individuals	25.1	23.3	3.4–78.0
	North Carolina Biotic Index (NCBI)	5.5	5.4	2.1–7.8

All metrics considered for inclusion in final indices were standardized on a 0 to 100 point scale. A Pearson product-moment correlation analysis was also performed on these metrics. Metrics with correlation values greater than 0.90 were not considered for inclusion in the same index. Metrics with correlation values greater than 0.80 (but less than 0.90) were plotted against each other to determine if the relationship was linear. Those metrics with correlation values greater than 0.80 and that have linear relationships were not considered for inclusion in the same index.

DISCRIMINATING INVERTEBRATE INDICES FOR THE CHARACTERISTIC REFERENCE STREAM

Invertebrate indices were developed for the Piedmont (Ecoregion 45) by combining different arrangements of discriminating metrics. Each candidate index included at least one metric from each of the metric categories (richness, composition, and tolerance/intolerance) that represent different aspects of invertebrate community composition. No functional feeding group or habit metric examined were found to differentiate reference from impaired communities. The index that best discriminated between reference sites and impaired sites is indicated next:

Index 45 (DE = 69%)
EPT Taxa
% Chironomidae
% Plecoptera
% Odonata
% EPT
NCBI

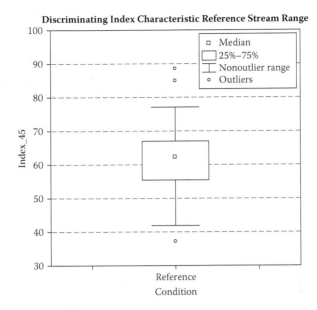

FIGURE E.2 Discriminating index characteristic between reference stream range in the Piedmont (Ecoregion 45).

SUBECOREGION 45B—SOUTHERN OUTER PIEDMONT: CHARACTERISTIC REFERENCE STREAM CONDITION

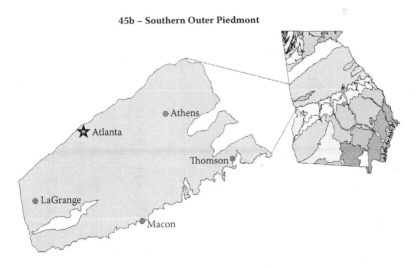

FIGURE E.3 Subecoregion 45b—Southern Outer Piedmont.

The Southern Outer Piedmont has lower elevations, less relief, and less precipitation than Subecoregion 45a. Loblolly-shortleaf pine is the major forest type, with less oak-hickory and oak-pine than in 45a. Gneiss, schist, and granite are the dominant rock types, covered with deep saprolite and mostly red, clayey subsoils. The majority of soils are Kanhapludults. The southern boundary of the subecoregion occurs at the Fall Line, where unconsolidated coastal plain sediments are deposited over the Piedmont metamorphic and igneous rocks.

TABLE E.3
Characteristic Reference Stream Land Use, Habitat, and Chemistry Data for Subecoregion 45b—Southern Outer Piedmont

	Parameter	Mean	Median	Range
Catchment Land Use	% Natural	86.7	86.8	80.2–92.9
	% Agriculture	0.90	0.50	0.1–2.3
	% Silviculture	7.0	5.2	3.0–14.7
	% Urban	5.4	5.6	4.0–6.5
Habitat	Total Habitat Score (200)	148.8	153.0	128–161
	Epifaunal Substrate (20)	14.6	15.0	8–19
	Embeddedness or Pool Substrate Characterization (20)	12.0	13.0	5–16
	Velocity/Depth Regime or Pool Variability (20)	15.6	16.0	13–17
	Sediment Deposition (20)	11.2	13.0	6–15
	Channel Flow Status (20)	15.8	17.0	13–18
	Channel Alteration (20)	17.0	17.0	15–19
	Frequency of Riffles or Channel Sinuosity (20)	14.6	15.0	12–16
	Bank Stability (L) (10)	6.8	7.0	6–8
	Bank Stability (R) (10)	7.0	7.0	5–9
	Vegetative Protection (L) (10)	8.0	8.0	7–9
	Vegetative Protection (R) (10)	8.4	8.5	8–9
	Riparian Vegetative Width (L) (10)	9.2	10.0	7–10
	Riparian Vegetative Width (R) (10)	8.6	9.0	5–10
In-Stream Habitat (substrate)	% Silt/Clay	9.4	1.9	0–42.0
	% Sand	56.1	55.3	24.8–89.0
	% Gravel	23.9	29.1	2.0–43.7
	% Cobble	6.5	1.0	0–22.8
	% Boulder	1.2	1.0	0–2.9
	% Bedrock	4.0	3.0	0–9.9
Chemistry (in situ)	Specific Conductivity (mS/cm)	0.081	0.082	0.044–0.106
	Dissolved Oxygen (mg/l)	9.7	7.8	7.2–13.8
	pH (SU)	6.9	6.8	6.6–7.2
	Turbidity (NTU)	10.0	10.5	0–20.3
Chemistry (laboratory)	Alkalinity (mg/l as $CaCO_3$)	24.9	32.5	0–44.9
	Total Hardness (mg/l as $CaCO_3$)	27.7	26.6	13.2–40.2
	Ammonia (mg/l as N)	0.36	0.07	BD–0.97
	Nitrate–Nitrite (mg/l as N)	0.08	0.09	0.01–0.16

(*Continued*)

TABLE E.3 (CONTINUED)
Characteristic Reference Stream Land Use, Habitat, and Chemistry Data for Subecoregion 45b—Southern Outer Piedmont

Parameter	Mean	Median	Range
Total Phosphorous (mg/l as P)	0.042	0.042	BD–0.042
Copper (mg/l)	0.004	0.004	BD–0.004
Iron (mg/l)	1.12	1.12	BD–1.12
Manganese (mg/l)	0.11	0.11	BD–0.11
Zinc (mg/l)	0.01	0.01	BD–0.01

Note: BD = Below detection.

DISCRIMINATING INVERTEBRATE METRICS FOR THE CHARACTERISTIC REFERENCE STREAM

Approximately 70 invertebrate metrics were evaluated for Subecoregion 45b based upon data collected from reference streams. Table E.4 includes raw data values for metrics that were judged by the strength of their discrimination between reference and impaired sites, and by graphical analysis, to be candidates for inclusion in final indices.

TABLE E.4
Discriminating Invertebrate Metrics for the Characteristic Reference Steam for Subecoregion 45b—Southern Outer Piedmont

Metric Category	Metric	Mean	Median	Range
Richness	Ephemeroptera, Plecoptera, Trichoptera Taxa (EPT Taxa)	14.4	14.0	9–20
	Trichoptera Taxa	5.5	5.0	3–9
	Coleoptera Taxa	6.8	5.0	3–9
Composition	% EPT	37.2	42.5	17–53
	% Chironomidae	29.2	27.1	22–36
	% Diptera	39.9	38.1	35–44
	% Plecoptera	9.3	10.8	1–17
	% Trichoptera	12.9	10.4	4–30
	% Oligochaeta	1.1	0.8	0–8
	% Coleoptera	9.6	8.8	7–15
Tolerance/ Intolerance	Hilsenhoff's Biotic Index (HBI)	5.9	5.9	4.8–6.9
	% Intolerant Individuals	11.9	7.6	5.8–23.3
Functional Feeding Group	% Shredder	13.8	12.1	6–22
	Scraper Taxa	5.8	6.0	2–9
Habit	Clinger Taxa	12.8	15.0	5–18
	Swimmer Taxa	2.8	3.0	1–4

All metrics considered for inclusion in final indices were standardized on a 0 to 100 point scale. A Pearson product-moment correlation analysis was also performed on these metrics. Metrics with correlation values greater than 0.90 were not considered for inclusion in the same index. Metrics with correlation values greater than 0.80 (but less than 0.90) were plotted against each other to determine if the relationship was linear. Those metrics with correlation values greater than 0.80 and that have linear relationships were not considered for inclusion in the same index.

DISCRIMINATING INVERTEBRATE INDICES FOR THE CHARACTERISTIC REFERENCE STREAM

Invertebrate indices were developed for Subecoregion 45b by combining different arrangements of discriminating metrics. Each candidate index included at least one metric from each of the metric categories (richness, composition, tolerance/intolerance, functional feeding group, and habit) that represent different aspects of invertebrate community composition. The index that best discriminated between reference sites and impaired sites is indicated next.

Index 45b (DE = 100%)
　　Coleoptera Taxa
　　% Oligochaeta
　　% Chironomidae
　　% Intolerant Individuals
　　Scraper Taxa
　　Swimmer Taxa

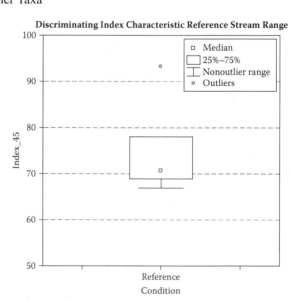

FIGURE E.4 Discriminating index characteristic reference stream range for Subecoregion 45b—Southern Outer Piedmont.

FIGURE E.5 Typical reference stream for Subecoregion 45b—Southern Outer Piedmont.

ECOREGION 65—SOUTHEASTERN PLAINS: CHARACTERISTIC REFERENCE STREAM CONDITION

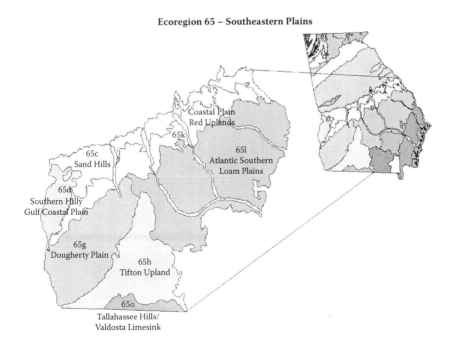

FIGURE E.6 Ecoregion 65—Southeastern Plains.

The Southeastern Plains are irregular plains with broad interstream areas that have a mosaic of cropland, pasture, woodland, and forest. Natural vegetation is mostly oak-hickory-pine and Southern mixed forest. The Cretaceous or Tertiary-age sands, silts, and clays of the region contrast geologically with the Paleozoic limestone, shale, and sandstone of Ecoregions 67 and 68, or with the even older metamorphic and igneous rocks of the Piedmont (Ecoregion 45). Elevations and relief are greater than in the Southern Coastal Plain (Ecoregion 75), but generally less than in much of the Piedmont. Streams in this area are relatively low gradient and sandy bottomed.

TABLE E.5
Characteristic Reference Stream Land Use, Habitat, and Chemistry Data for Ecoregion 65—Southeastern Plains

	Parameter	Mean	Median	Range
Catchment	% Natural	71.3	73.9	43.3–94.8
Land Use	% Agriculture	22.5	21.2	0.7–51.2
	% Silviculture	7.8	6.4	0.2–24.8
	% Urban	5.7	5.5	0.7–11.8
Habitat	Total Habitat Score (200)	158.2	161.5	121–179
	Epifaunal Substrate (20)	15.6	16.0	11–18
	Pool Substrate Characterization (20)	14.3	15.0	7–18
	Pool Variability (20)	14.7	16.0	7–19
	Sediment Deposition (20)	15.3	16.0	6–18
	Channel Flow Status (20)	16.5	17.0	10–19
	Channel Alteration (20)	17.2	17.0	15–20
	Channel Sinuosity (20)	14.0	14.5	5–20
	Bank Stability (L) (10)	8.3	9.0	4–10
	Bank Stability (R) (10)	8.3	9.0	4–10
	Vegetative Protection (L) (10)	8.9	9.0	7–10
	Vegetative Protection (R) (10)	9.1	9.0	6–10
	Riparian Vegetative Width (L) (10)	8.0	8.0	4–10
	Riparian Vegetative Width (R) (10)	8.0	8.0	4–10
In-Stream Habitat (substrate)	% Silt/Clay	17.5	7.0	0.0–100.0
	% Sand	79.7	90.5	0.0–100.0
	% Gravel	2.7	0.0	0.0–30.5
	% Cobble	0.6	0.0	0.0–16.7
	% Boulder	0.1	0.0	0.0–1.9
	% Bedrock	0.1	0.0	0.0–2.0
Chemistry (in situ)	Specific Conductivity (mS/cm)	0.097	0.060	0.036–0.089
	Dissolved Oxygen (mg/l)	9.3	9.3	5.5–16.5
	pH (SU)	6.1	6.3	4.1–7.5
	Turbidity (NTU)	8.3	6.9	0.0–39.6

(Continued)

TABLE E.5 (CONTINUED)
Characteristic Reference Stream Land Use, Habitat, and Chemistry Data for Ecoregion 65—Southeastern Plains

	Parameter	Mean	Median	Range
Chemistry (laboratory)	Alkalinity (mg/l as $CaCO_3$)	23.2	8.6	0.0–176.0
	Total Hardness (mg/l as $CaCO_3$)	37.0	21.2	0.0–196.9
	Ammonia (mg/l as N)	0.056	0.054	BD–0.089
	Nitrate–Nitrite (mg/l as N)	0.160	0.076	BD–0.806
	Total Phosphorous (mg/l as P)	0.085	0.054	BD–0.209
	Copper (mg/l)	0.003	0.003	BD–0.003
	Iron (mg/l)	1.98	0.82	BD–12.99
	Manganese (mg/l)	0.092	0.092	BD–0.141
	Zinc (mg/l)	0.036	0.034	BD–0.052

Note: BD = Below detection.

DISCRIMINATING INVERTEBRATE METRICS FOR THE CHARACTERISTIC REFERENCE STREAM

Approximately 70 invertebrate metrics were evaluated for Ecoregion 65 based upon data collected from reference streams. Table E.6 includes raw data values for metrics that were judged by the strength of their discrimination between reference and impaired sites, and by graphical analysis, to be candidates for inclusion in final indices.

TABLE E.6
Discriminating Invertebrate Metrics for the Characteristic Reference Stream for Ecoregion 65—Southeastern Plains

Metric Category	Metric	Mean	Median	Range
Richness	Ephemeroptera, Plecoptera, Trichoptera Taxa (EPT Taxa)	6.6	6.5	0–16
	Margalef's Index	7.6	7.8	2.1–12.0
	Total Taxa	41.8	42.0	12–67
Composition	% Oligochaeta	2.3	1.4	0–11.7
	% Tanypodinae/Total Chironomidae (TC)	17.1	15.2	0–61.0
Tolerance/ Intolerance	% Intolerant Individuals	11.1	6.4	0–46.7
	Beck's Index	8.9	8.0	2–21
Functional Feeding Group	% Predator	15.1	11.8	1.5–48.8
Habit	% Clinger	19.3	18.8	0–63.3

All metrics considered for inclusion in final indices were standardized on a 0 to 100 point scale. A Pearson product-moment correlation analysis was also performed on these metrics. Metrics with correlation values greater than 0.90 were not considered for inclusion in the same index. Metrics with correlation values greater than 0.80 (but less than 0.90) were plotted against each other to determine if the relationship was linear. Those metrics with correlation values greater than 0.80 and that have linear relationships were not considered for inclusion in the same index.

DISCRIMINATING INVERTEBRATE INDICES FOR THE CHARACTERISTIC REFERENCE STREAM

Invertebrate indices were developed for the Southeastern Plains (Ecoregion 65) by combining different arrangements of discriminating metrics. Each candidate index included at least one metric from each of the metric categories (richness, composition, tolerance/intolerance, functional feeding group, and habit) that represent different aspects of invertebrate community composition. The index that best discriminated between reference sites and impaired sites is indicated next.

Index 65 (DE = 62%)
    EPT Taxa
    Margalef's Index
    % Oligochaeta
    % Tanypodinae/TC
    % Intolerant Individuals
    % Predator
    % Clinger

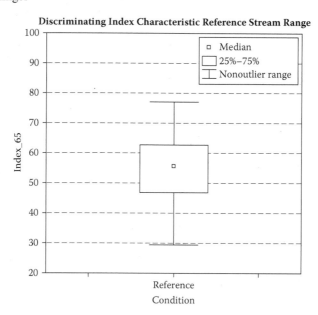

FIGURE E.7   Discriminating index characteristic reference stream range for Ecoregion 65.

SUBECOREGION 65C—SAND HILLS: CHARACTERISTIC REFERENCE STREAM CONDITION

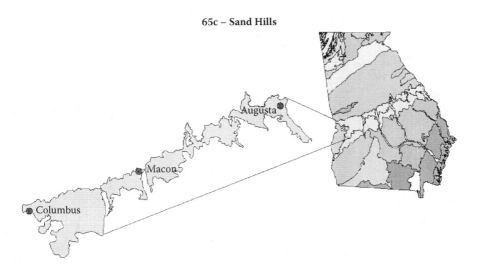

65c – Sand Hills

FIGURE E.8   Subecoregion 65c—Sand Hills.

The Sand Hills of Georgia form a narrow, rolling to hilly, highly dissected coastal plain belt stretching across the state from Augusta to Columbus. The region is composed primarily of Cretaceous and some Eocene-age marine sands and clays deposited over the crystalline and metamorphic rocks of the Piedmont (Ecoregion 45). Many of the droughty, low-nutrient soils formed in thick beds of sand, although soils in some areas contain more loamy and clayey horizons. On the drier sites, turkey oak and longleaf pine are dominant, while shortleaf-loblolly pine forests and other oak-pine forests are common throughout the region.

TABLE E.7
Characteristic Reference Stream Land Use, Habitat, and Chemistry Data for Subecoregion 65c—Sand Hills

	Parameter	Mean	Median	Range
Catchment	% Natural	72.5	72.2	65.4–77.5
Land Use	% Agriculture	7.1	8.4	0–13.1
	% Silviculture	15.3	15.3	9.0–21.1
	% Urban	5.1	5.2	3.0–7.3
Habitat	Total Habitat Score (200)	164.4	164.0	159–170
	Epifaunal Substrate (20)	15.6	16.0	13–18
	Pool Substrate Characterization (20)	13.8	15.0	9–16
	Pool Variability (20)	14.8	16.0	10–16
	Sediment Deposition (20)	17.0	17.0	16–18
	Channel Flow Status (20)	19.0	19.0	19
	Channel Alteration (20)	18.4	19.0	17–19
	Channel Sinuosity (20)	11.8	13.0	9–15
	Bank Stability (L) (10)	8.8	9.0	8–9
	Bank Stability (R) (10)	9.2	9.0	8–10
	Vegetative Protection (L) (10)	8.4	8.0	8–9
	Vegetative Protection (R) (10)	8.4	8.0	8–9
	Riparian Vegetative Width (L) (10)	9.4	10.0	8–10
	Riparian Vegetative Width (R) (10)	9.8	10.0	9–10
In-Stream	% Silt/Clay	37.0	12.0	0–22.8
Habitat	% Sand	87.0	95.7	63.0–100.0
(substrate)	% Gravel	1.1	0	0–4.3
	% Cobble	0	0	0
	% Boulder	0	0	0
	% Bedrock	0	0	0
Chemistry	Specific Conductivity (mS/cm)	0.020	0.015	0.003–0.049
(in situ)	Dissolved Oxygen (mg/l)	11.3	11.7	10.3–12.5
	pH (SU)	5.1	5.1	4.3–6.2
	Turbidity (NTU)	2.3	1.1	0–6.9
Chemistry	Alkalinity (mg/l as $CaCO_3$)	1.8	0	0–8.2
(laboratory)	Total Hardness (mg/l as $CaCO_3$)	9.8	10.3	5.5–18.0
	Ammonia (mg/l as N)	0.054	0.052	BD–0.07
	Nitrate–Nitrite (mg/l as N)	0.18	0.11	0.07–0.47
	Total Phosphorous (mg/l as P)	BD	BD	BD
	Copper (mg/l)	BD	BD	BD
	Iron (mg/l)	0.54	0.54	BD–0.92
	Manganese (mg/l)	BD	BD	BD
	Zinc (mg/l)	BD	BD	BD

Note: BD = Below detection.

DISCRIMINATING INVERTEBRATE METRICS
FOR THE CHARACTERISTIC REFERENCE STREAM

Approximately 70 invertebrate metrics were evaluated for Subecoregion 65c based upon data collected from reference sites. Table E.8 includes raw data values for metrics that were judged by the strength of their discrimination of reference from stressor sites, and by graphical analysis, to be candidates for inclusion in final indices.

TABLE E.8
Discriminating Invertebrate Metrics for the Characteristic Reference Stream for Subecoregion 65c—Sand Hills

Metric Category	Metric	Mean	Median	Range
Richness	Trichoptera Taxa	6.0	7.0	3–8
	Plecoptera Taxa	2.2	2.0	1–4
Composition	% Trichoptera	11.7	8.8	4–26
Tolerance/Intolerance	% *Cricotopus* and *Chironomus*/Total Chironomidae (TC)	0	0	0
	Beck's Biotic Index	14.6	14	8–21
Functional Feeding Group	% Scraper	15.5	11.3	3.6–28.0
	% Predator	25.2	19.6	10–39
Habit	Clinger Taxa	13.0	12	10–17

All metrics considered for inclusion in final indices were standardized on a 0 to 100 point scale. A Pearson product-moment correlation analysis was also performed on these metrics. Metrics with correlation values greater than 0.90 were not considered for inclusion in the same index. Metrics with correlation values greater than 0.80 (but less than 0.90) were plotted against each other to determine if the relationship was linear. Those metrics with correlation values greater than 0.80 and that have linear relationships were not considered for inclusion in the same index.

DISCRIMINATING INVERTEBRATE INDICES
FOR THE CHARACTERISTIC REFERENCE STREAM

Invertebrate indices were developed for Subecoregion 65c by combining different arrangements of discriminating metrics. Each candidate index included at least one metric from each of the metric categories (richness, composition, tolerance/intolerance, functional feeding group, and habit) that represent different aspects of invertebrate community composition. The index that best discriminated between reference sites and impaired sites is indicated next.

Index 65c (DE = 86%)
  Plecoptera Taxa
  % Plecoptera
  % Trichoptera
  Cricotopus and *Chironomus*/Total Chironomidae
  Scraper Taxa
  Clinger Taxa

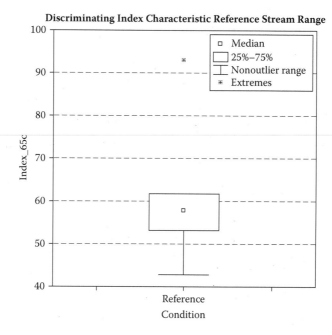

FIGURE E.9 Discriminating index characteristics reference stream range for Subecoregion 65c—Sand Hills.

FIGURE E.10   Typical reference stream for Subecoregion 65c—Sand Hills.

ECOREGION 66—BLUE RIDGE: CHARACTERISTIC REFERENCE STREAM CONDITION

Ecoregion 66 – Blue Ridge

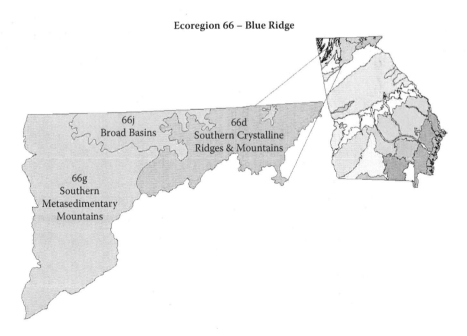

FIGURE E.11   Ecoregion 66—Blue Ridge.

The Blue Ridge extends from southern Pennsylvania to northern Georgia, varying from narrow ridges to hilly plateaus to more massive mountainous areas with high peaks. The mostly forested slopes; high-gradient, cool, clear streams; and rugged terrain occur on a mix of igneous, metamorphic, and sedimentary geology. Annual precipitation of over 80 inches can occur on the well-exposed high peaks. The southern Blue Ridge is one of the richest centers of biodiversity in the eastern United States. It is one of the most floristically diverse ecoregions, and includes Appalachian oak forests, northern hardwoods, and, at the highest elevations in Tennessee and North Carolina, Southeastern spruce-fir forests. Shrub, grass, heath balds, hemlock, cove hardwoods, and oak-pine communities are also significant. Black bear, whitetail deer, wild boar, turkey, grouse, songbirds, many species of amphibians and reptiles, thousands of species of invertebrates, and a variety of small mammals are found here.

TABLE E.9

Characteristic Reference Stream Land Use, Habitat, and Chemistry Data for Ecoregion 66—Blue Ridge

	Parameter	Mean	Median	Range
Catchment	% Natural	91.3	95.9	61.6–99.8
Land Use	% Agriculture	2.9	0.2	0.0–11.7
	% Silviculture	0.4	0.1	0.0–4.0
	% Urban	2.4	1.9	0.0–6.4
Habitat	Total Habitat Score (200)	166.7	167.5	111–192
	Epifaunal Substrate (20)	16.1	16.5	10–19
	Embeddedness (20)	16.0	17.5	4–19
	Velocity/Depth Regime (20)	16.6	17.0	14–19
	Sediment Deposition (20)	15.6	17.0	5–19
	Channel Flow Status (20)	16.5	16.5	13–20
	Channel Alteration (20)	18.1	18.0	15–20
	Frequency of Riffles (20)	17.9	18.0	16–20
	Bank Stability (L) (10)	8.8	9.0	4–10
	Bank Stability (R) (10)	8.5	9.0	3–10
	Vegetative Protection (L) (10)	8.4	9.0	3–10
	Vegetative Protection (R) (10)	7.9	9.0	3–10
	Riparian Vegetative Width (L) (10)	8.9	10.0	1–10
	Riparian Vegetative Width (R) (10)	7.6	8.5	1–10
In-Stream	% Silt/clay	2.6	0.0	0.0–12.1
Habitat	% Sand	8.9	7.0	0.0–28.0
(substrate)	% Gravel	37.5	35.0	13.2–69.0
	% Cobble	34.3	32.4	4.0–54.0
	% Boulder	14.8	14.0	0.0–33.0
	% Bedrock	1.9	0.0	0.0–8.0
Chemistry	Specific Conductivity (mS/cm)	0.017	0.016	0.008–0.038
(*in situ*)	Dissolved Oxygen (mg/l)	11.0	10.9	8.9–13.0
	pH (SU)	6.8	6.8	6.4–7.2
	Turbidity (NTU)	5.0	4.7	0.0–17.8

(Continued)

TABLE E.9 (CONTINUED)
Characteristic Reference Stream Land Use, Habitat, and Chemistry Data for Ecoregion 66—Blue Ridge

	Parameter	Mean	Median	Range
Chemistry	Alkalinity (mg/l as CaCO₃)	6.1	6.1	0.0–12.3
(laboratory)	Total Hardness (mg/l as CaCO₃)	6.6	6.5	2.7–15.4
	Ammonia (mg/l as N)	0.049	0.046	BD–0.036
	Nitrate–Nitrite (mg/l as N)	0.186	0.085	BD–0.841
	Total Phosphorous (mg/l as P)	0.089	0.075	BD–0.062
	Copper (mg/l)	BD	BD	BD
	Iron (mg/l)	0.151	0.102	BD–0.458
	Manganese (mg/l)	0.010	0.006	BD–0.029
	Zinc (mg/l)	0.011	0.012	BD–0.006

Note: BD = Below detection.

DISCRIMINATING INVERTEBRATE METRICS FOR THE CHARACTERISTIC REFERENCE STREAM

Approximately 70 invertebrate metrics were evaluated for Ecoregion 66 based upon data collected from reference streams. Table E.10 includes raw data values for metrics that were judged by the strength of their discrimination between reference and impaired sites, and by graphical analysis, to be candidates for inclusion in final indices.

TABLE E.10
Discriminating Invertebrate Metrics for the Characteristic Reference Stream for Ecoregion 66—Blue Ridge

Metric Category	Metric	Mean	Median	Range
Richness	Ephemeroptera, Plecoptera, Trichoptera Taxa (EPT Taxa)	24.4	25.0	11–37
	Simpson's Index	0.042	0.038	0.07–0.02
	Plecoptera Taxa	7.7	9.0	3–12
Composition	% Trichoptera	19.6	20.0	9.6–26.3
Tolerance/	Beck's Index	37.2	39.0	19–51
Intolerance	% Intolerant Individuals	36.8	37.9	12.5–54.7
	North Carolina Biotic Index (NCBI)	4.2	4.2	3.3–5.5
Functional Feeding	Predator Taxa	13.3	13.0	8–19
Group	Shredder Taxa	8.0	8.0	1–15
Habit	Clinger Taxa	22.1	22.0	12–32
	Burrower Taxa	7.2	7.0	4–11

All metrics considered for inclusion in final indices were standardized on a 0 to 100 point scale. A Pearson product-moment correlation analysis was also performed on these metrics. Metrics with correlation values greater than 0.90 were not considered for inclusion in the same index. Metrics with correlation values greater than 0.80 (but less than 0.90) were plotted against each other to determine if the relationship was linear. Those metrics with correlation values greater than 0.80 and that have linear relationships were not considered for inclusion in the same index.

DISCRIMINATING INVERTEBRATE INDICES FOR THE CHARACTERISTIC REFERENCE STREAM

Invertebrate indices were developed for the Blue Ridge (Ecoregion 66) by combining different arrangements of discriminating metrics. Each candidate index included at least one metric from each of the metric categories (richness, composition, tolerance/intolerance, functional feeding group, and habit) that represent different aspects of invertebrate community composition. The index that best discriminated between reference sites and impaired sites is indicated next.

Index 66 (DE = 76%)
    Plecoptera Taxa
    Simpson's Index
    % Trichoptera
    % Intolerant Individuals
    NCBI
    Predator Taxa
    Burrower Taxa

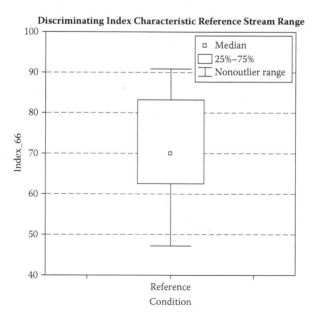

FIGURE E.12 Discriminating index characteristic reference stream range for Ecoregion 66—Blue Ridge.

SUBECOREGION 66D—SOUTHERN CRYSTALLINE RIDGES AND MOUNTAINS: CHARACTERISTIC SUBECOREGION REFERENCE STREAM CONDITION

66d – Southern Crystalline Ridges and Mountains

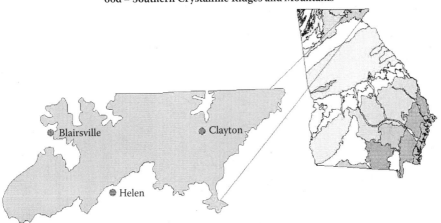

FIGURE E.13   Subecoregion 66d—Southern Crystalline Ridges and Mountains.

The Southern Crystalline Ridges and Mountains contain the highest and wettest mountains in Georgia. These occur primarily on Precambrian-age igneous and high-grade metamorphic rocks. The common crystalline rock types include gneiss, schist, and quartzite, covered by well-drained, acidic, brownish, loamy soils. Some mafic and ultramafic rocks also occur here, producing more basic soils. Elevations of this rough, dissected region are typically 1800 to 4000 feet, with Brasstown Bald Mountain, the highest point in Georgia, reaching 4784 feet. Although there are a few small areas of pasture and apple orchards, the region is mostly forested.

TABLE E.11
Characteristic Reference Stream Land Use, Habitat, and Chemistry Data for Subecoregion 66d—Southern Crystalline Ridges and Mountains

	Parameter	Mean	Median	Range
Catchment	% Natural	97.7	97.9	97.2–98.1
Land Use	% Agriculture	0	0	0
	% Silviculture	0.30	0.30	0.10–0.50
	% Urban	1.9	1.9	1.5–2.6
Habitat	Total Habitat Score (200)	174.6	179.0	162–186
	Epifaunal Substrate (20)	16.8	17.0	14–19
	Embeddedness (20)	16.6	17.0	15–18
	Velocity/Depth Regime (20)	17.6	18.0	15–19
	Sediment Deposition (20)	16.0	16.0	13–18
	Channel Flow Status (20)	16.8	17.0	15–18
	Channel Alteration (20)	18.2	18.0	17–20
	Frequency of Riffles (20)	18.2	19.0	16–19
	Bank Stability (L) (10)	9.2	9.0	8–10
	Bank Stability (R) (10)	9.0	9.0	8–10
	Vegetative Protection (L) (10)	9.2	9.0	8–10
	Vegetative Protection (R) (10)	8.4	9.0	7–9
	Riparian Vegetative Width (L) (10)	9.8	10.0	9–10
	Riparian Vegetative Width (R) (10)	8.8	9.0	7–10
In-Stream	% Silt/Clay	0.2	0	0–1.0
Habitat	% Sand	6.2	6.0	2.9–8.0
(substrate)	% Gravel	33.1	32.0	28.4–40.2
	% Cobble	39.3	42.0	28.4–54.0
	% Boulder	17.4	17.6	6.0–30.0
	% Bedrock	3.8	3.0	0–8.0
Chemistry	Specific Conductivity (mS/cm)	0.013	0.012	0.008–0.016
(in situ)	Dissolved Oxygen (mg/l)	11.2	11.6	9.7–11.9
	pH (SU)	6.7	6.6	6.4–7.1
	Turbidity (NTU)	1.42	0.5	0–5.8
Chemistry	Alkalinity (mg/l as $CaCO_3$)	5.6	5.5	2.5–8.3
(laboratory)	Total Hardness (mg/l as $CaCO_3$)	6.3	4.0	3.5–10.4
	Ammonia (mg/l as N)	0.05	0.051	0.037–0.057
	Nitrate–Nitrite (mg/l as N)	0.052	0.063	BD–0.07
	Total Phosphorous (mg/l as P)	0.142	0.142	BD–0.142
	Copper (mg/l)	BD	BD	BD
	Iron (mg/l)	0.04	0.04	BD–0.04
	Manganese (mg/l)	0.006	0.006	BD–0.006
	Zinc (mg/l)	0.006	0.006	BD–0.006

Note: BD = Below detection.

Discriminating Invertebrate Metrics for the Characteristic Reference Stream

Approximately 70 invertebrate metrics were evaluated for Subecoregion 66d based upon data collected from reference sites. Table E.12 includes raw data values for metrics that were judged by the strength of their discrimination of reference from stressor sites, and by graphical analysis, to be candidates for inclusion in final indices.

TABLE E.12
Discriminating Invertebrate Metrics for the Characteristic Reference Stream for Subecoregion 66d—Southern Crystalline Ridges and Mountains

Metric Category	Metric	Mean	Median	Range
Richness	Ephemeroptera, Plecoptera, and Trichoptera Taxa (EPT)	23.0	24.0	12–31
	Diptera Taxa	24.2	25.0	16–31
Composition	% Plecoptera	22.5	24.7	11.3–30.8
	% Odonata	1.2	0.4	0–4.6
Tolerance/ Intolerance	Intolerant Taxa	6.6	6.0	4–10
	% Dominant Individuals	26.2	26.0	18–38
Functional Feeding Group	% Shredder	20.5	14.2	8–34
	Predator Taxa	14.6	13.0	12–19
Habit	Clinger Taxa	22.2	22.0	15–32

All metrics considered for inclusion in final indices were standardized on a 0 to 100 point scale. A Pearson product-moment correlation analysis was also performed on these metrics. Metrics with correlation values greater than 0.90 were not considered for inclusion in the same index. Metrics with correlation values greater than 0.80 (but less than 0.90) were plotted against each other to determine if the relationship was linear. Those metrics with correlation values greater than 0.80 and that have linear relationships were not considered for inclusion in the same index.

Discriminating Invertebrate Indices for the Characteristic Reference Stream

Invertebrate indices were developed for Subecoregion 66d by combining different arrangements of discriminating metrics. Each candidate index included at least one metric from each of the metric categories (richness, composition, tolerance/ intolerance, functional feeding group, and habit) that represent different aspects of invertebrate community composition. The index that best discriminated between reference sites and impaired sites is indicated next.

Index 66d (DE = 80%)
 Diptera Taxa
 % Plecoptera
 % Odonata
 % Dominant Individuals
 % Shredder
 Clinger Taxa

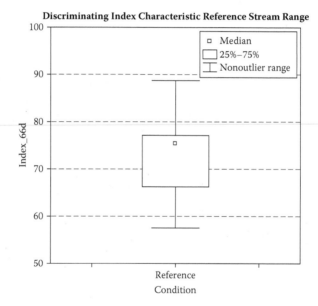

FIGURE E.14 Discriminating index characteristic reference stream range for Subecoregion 66d—Southern Crystalline Ridges and Mountains.

FIGURE E.15 Typical reference stream for Subecoregion 66d–Southern Crystalline Ridges and Mountains.

ECOREGION 75—SOUTHERN COASTAL PLAIN: CHARACTERISTIC ECOREGION REFERENCE STREAM CONDITION

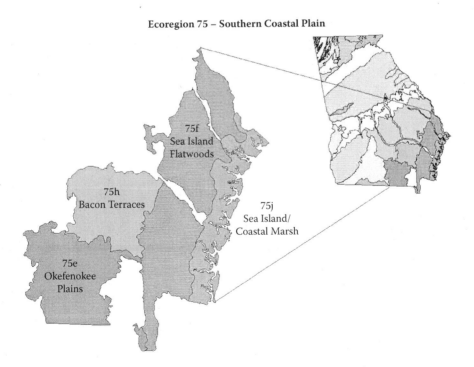

Ecoregion 75 – Southern Coastal Plain

75f
Sea Island
Flatwoods

75h
Bacon Terraces

75j
Sea Island/
Coastal Marsh

75e
Okefenokee
Plains

FIGURE E.16 Ecoregion 75—Southern Coastal Plain.

The Southern Coastal Plain extends from South Carolina and Georgia through much of central Florida, and along the Gulf coast lowlands of the Florida Panhandle, Alabama, and Mississippi. From a national perspective, it appears to be mostly flat plains, but it is a heterogeneous region also containing barrier islands, coastal lagoons, marshes, and swampy lowlands along the Gulf and Atlantic coasts. In Florida, an area of discontinuous highlands contains numerous lakes. This ecoregion is generally lower in elevation with less relief and wetter soils than Ecoregion 65. Once covered by a variety of forest communities that included trees of longleaf pine, slash pine, pond pine, beech, sweetgum, southern magnolia, white oak, and laurel oak, land cover in the region is now mostly slash and loblolly pine with oak-gum-cypress forest in some low lying areas, citrus groves, pasture for beef cattle, and urban.

TABLE E.13
Characteristic Reference Stream Land Use, Habitat, and Chemistry Data for Ecoregion 75—Southern Coastal Plain

	Parameter	Mean	Median	Range
Catchment	% Natural	87.0	90.5	64.5–95.9
Land Use	% Agriculture	4.9	1.5	0.0–27.1
	% Silviculture	11.8	11.3	0.2–35.0
	% Urban	8.0	6.5	4.1–20.5
Habitat	Total Habitat Score (200)	152.2	152.0	112–181
	Epifaunal Substrate (20)	14.8	15.0	10–19
	Pool Substrate Characterization (20)	13.0	13.0	8–19
	Pool Variability (20)	11.7	11.0	5–19
	Sediment Deposition (20)	15.8	17.5	8–20
	Channel Flow Status (20)	15.8	18.0	5–20
	Channel Alteration (20)	18.2	18.0	13–20
	Channel Sinuosity (20)	13.6	13.0	8–20
	Bank Stability (L) (10)	8.3	9.0	1–10
	Bank Stability (R) (10)	8.3	9.0	1–10
	Vegetative Protection (L) (10)	8.0	8.0	3–10
	Vegetative Protection (R) (10)	8.1	8.5	3–10
	Riparian Vegetative Width (L) (10)	8.6	9.0	5–10
	Riparian Vegetative Width (R) (10)	8.0	9.0	3–10
In-Stream	% Silt/clay	28.2	12.9	0.0–100.0
Habitat	% Sand	71.4	85.6	0.0–100.0
(substrate)	% Gravel	0.4	0.0	0.0–4.0
	% Cobble	0.0	0.0	0.0
	% Boulder	0.0	0.0	0.0
	% Bedrock	0.0	0.0	0.0
Chemistry	Specific Conductivity (mS/cm)	0.871	0.108	0.051–8.920
(in situ)	Dissolved Oxygen (mg/l)	6.7	6.6	3.5–14.6
	pH (SU)	4.8	4.5	3.6–6.7
	Turbidity (NTU)	11.5	6.7	0.0–57.0

(Continued)

TABLE E.13 (CONTINUED)
Characteristic Reference Stream Land Use, Habitat, and Chemistry Data for Ecoregion 75—Southern Coastal Plain

	Parameter	Mean	Median	Range
Chemistry	Alkalinity (mg/l as CaCO₃)	8.8	0.0	0.0–101.4
(laboratory)	Total Hardness (mg/l as CaCO₃)	135.5	33.2	7.7–1067.0
	Ammonia (mg/l as N)	5,397	0.083	BD–48.917
	Nitrate–Nitrite (mg/l as N)	0.117	0.051	BD–0.325
	Total Phosphorous (mg/l as P)	0.138	0.122	BD–0.323
	Copper (mg/l)	0.009	0.009	BD–0.015
	Iron (mg/l)	1.076	1.015	BD–2.897
	Manganese (mg/l)	0.040	0.036	BD–0.099
	Zinc (mg/l)	0.018	0.017	BD–0.023

Note: BD = Below detection.

DISCRIMINATING INVERTEBRATE METRICS FOR THE CHARACTERISTIC REFERENCE STREAM

Approximately 70 invertebrate metrics were evaluated for Ecoregion 75 based upon data collected from reference streams. Table E.14 includes raw data values for metrics that were judged by the strength of their discrimination between reference and impaired sites, and by graphical analysis, to be candidates for inclusion in final indices.

TABLE E.14
Discriminating Invertebrate Metrics for the Characteristic Reference Stream for Ecoregion 75—Southern Coastal Plains

Metric Category	Metric	Mean	Median	Range
Composition	% Noninsect	25.4	16.7	0.5–92.4
	% Oligochaeta	2.2	1.0	0.0–8.1
	% Odonata	0.8	0.0	0.0–9.2
	% Tanypodinae/total Chironomidae (TC)	3.1	0.2	0.0–34.4
Tolerance/ Intolerance	Hilsenhoff's Biotic Index (HBI)	7.1	7.2	5.2–9.0

All metrics considered for inclusion in final indices were standardized on a 0 to 100 point scale. A Pearson product-moment correlation analysis was also performed on these metrics. Metrics with correlation values greater than 0.90 were not considered for inclusion in the same index. Metrics with correlation values greater than 0.80 (but less than 0.90) were plotted against each other to determine if the relationship was linear. Those metrics with correlation values greater than 0.80 and that have linear relationships were not considered for inclusion in the same index.

DISCRIMINATING INVERTEBRATE INDICES FOR THE CHARACTERISTIC REFERENCE STREAM

Invertebrate indices were developed for the Southern Coastal Plain (Ecoregion 75) by combining different arrangements of discriminating metrics. Each candidate index included at least one metric from each of the metric categories (composition and tolerance/intolerance) that represent different aspects of invertebrate community composition. No richness, functional feeding group, or habit metric examined was found to differentiate reference from impaired communities. The index that best discriminated between reference sites and impaired sites is indicated next.

Index 75 (DE = 77%)
% Noninsect
% Oligochaeta
% Odonata
% Tanypodinae/TC
HBI

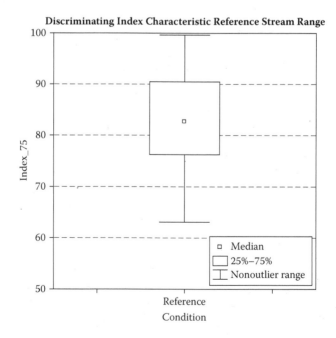

FIGURE E.17   Discriminating index characteristic reference stream range for Ecoregion 75—Southern Coastal Plain.

SUBECOREGION 75F—SEA ISLAND FLATWOODS: CHARACTERISTIC REFERENCE STREAM CONDITION

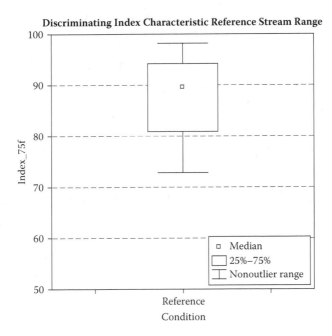

FIGURE E.18   Subecoregion 75f—Sea Island Flatwoods.

The Sea Island Flatwoods are poorly drained flat plains with lower elevations and less dissection than Subecoregion 65l. Pleistocene sea levels rose and fell several times creating different terraces and shoreline deposits. Spodosols and other wet soils are common, although small areas of better drained soils add some ecological diversity. Trail Ridge is in this region, forming the boundary with Subecoregion 75g. Loblolly and slash pine plantations cover much of the region. Water oak, willow oak, sweetgum, blackgum, and cypress occur in wet areas.

TABLE E.15
Characteristic Reference Stream Land Use, Habitat, and Chemistry Data for Subecoregion 75f—Sea Island Flatwoods

	Parameter	Mean	Median	Range
Catchment	% Natural	82.8	81.5	78.5–88.6
Land Use	% Agriculture	0	0	0
	% Silviculture	9.8	12.2	4.8–12.5
	% Urban	7.3	6.5	6.0–9.3
Habitat	Total Habitat Score (200)	153.0	151.0	146–164
	Epifaunal Substrate (20)	16.0	17.5	11–18
	Pool Substrate Characterization (20)	14.0	15.0	10–16
	Pool Variability (20)	11.0	10.0	9–15
	Sediment Deposition (20)	15.3	15.5	10–20
	Channel Flow Status (20)	18.0	19.5	13–20
	Channel Alteration (20)	17.5	17.5	15–20
	Channel Sinuosity (20)	12.3	10.5	9–19
	Bank Stability (L) (10)	8.8	8.5	8–10
	Bank Stability (R) (10)	8.8	8.5	8–10
	Vegetative Protection (L) (10)	8.0	8.5	6–9
	Vegetative Protection (R) (10)	8.3	8.5	6–10
	Riparian Vegetative Width (L) (10)	8.3	9.0	6–9
	Riparian Vegetative Width (R) (10)	7.0	8.0	3–9
In-Stream	% Silt/Clay	51.8	53.5	0–100
Habitat	% Sand	48.3	46.5	0–100.0
(substrate)	% Gravel	0	0	0
	% Cobble	0	0	0
	% Boulder	0	0	0
	% Bedrock	0	0	0
Chemistry	Conductivity (mS/cm)	.117	.120	.051–.179
(in situ)	Dissolved Oxygen (mg/l)	5.7	6.6	3.5–7.1
	pH (SU)	4.6	4.2	3.7–6.0
	Turbidity (NTU)	6.9	3.4	0–17.4
Chemistry	Alkalinity (mg/l as $CaCO_3$)	20.9	20.9	20.9
(laboratory)	Hardness (mg/l as $CaCO_3$)	40.1	40.1	40.1
	Ammonia (mg/l as N)	6.43	6.43	BD–6.43
	Nitrate–Nitrite (mg/l as N)	0.136	0.053	BD–0.315
	Total Phosphorous (mg/l as P)	0.089	0.089	BD–0.113
	Copper (mg/l)	0.003	0.003	BD–0.003
	Iron (mg/l)	0.87	0.96	BD–1.19
	Manganese (mg/l)	0.036	0.036	BD–0.036
	Zinc (mg/l)	0.017	0.017	BD–0.017

Note: BD = Below detection.

Discriminating Invertebrate Metrics for the Characteristic Reference Stream

Approximately 70 invertebrate metrics were evaluated for Subecoregion 75f based upon data collected from reference streams. Table E.16 includes raw data values for metrics that were judged by the strength of their discrimination between reference and impaired sites, and by graphical analysis, to be candidates for inclusion in final indices.

TABLE E.16
Discriminating Invertebrate Metrics for the Characteristic Reference Stream for Subecoregion 75f—Sea Island Flatwoods

Metric Category	Metric	Mean	Median	Range
Richness	Chironomidae Taxa	7.8	8.0	3–12
Composition	% Odonata	0.0	0.0	0.0
	% Amphipoda	2.4	0.8	0–8.1
	% Oligochaeta	3.0	1.9	0–8.1
	% Tanypodinae/Total Chironomidae (TC)	0.1	0.0	0–0.5
Tolerance/ Intolerance	Hilsenhoff's Biotic Index (HBI)	7.6	7.3	7.0–9.0
	Tolerant Taxa	6.8	6.5	4–10
Functional Feeding Group	% Filterer	0.3	0.2	0–0.9

All metrics considered for inclusion in final indices were standardized on a 0 to 100 point scale. A Pearson product-moment correlation analysis was also performed on these metrics. Metrics with correlation values greater than 0.90 were not considered for inclusion in the same index. Metrics with correlation values greater than 0.80 (but less than 0.90) were plotted against each other to determine if the relationship was linear. Those metrics with correlation values greater than 0.80 and that have linear relationships were not considered for inclusion in the same index.

Discriminating Invertebrate Indices for the Characteristic Reference Stream

Invertebrate indices were developed for Subecoregion 75f by combining different arrangements of discriminating metrics. Each candidate index included at least one metric from each of the metric categories (richness, composition, tolerance/intolerance, and functional feeding group) that represent different aspects of invertebrate community composition. No habit metric examined was found to differentiate reference from impaired communities. The index that best discriminated between reference sites and impaired sites is indicated next.

Index 75f (DE = 100%)
   Chironomidae Taxa
   % Odonata
   % Oligochaeta
   % Tanypodinae/TC
   Tolerant Taxa
   % Filterer

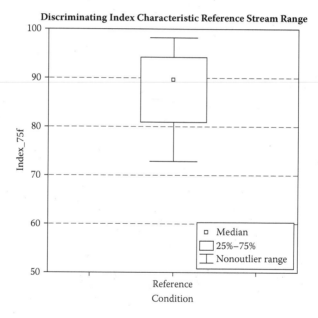

FIGURE E.19 Discriminating index characteristic reference stream range for Subecoregion 75f—Sea Island Flatwoods.

FIGURE E.20   Typical reference stream for Subecoregion 75f.

Appendix F: Examples of Reference Criteria, Numerical Rating Systems, and Discrimination Efficiencies from the State of Georgia

CONTENTS

The following text describes the appropriate macroinvertebrate metric index for sample ecoregions or subecoregions, the descriptive statistics for each of the metrics (that is, the range of that metric for each of the five classifications), a depiction of the ability of the index to discriminate between reference and impaired conditions, the suggested range of scores for the numeric ranking, and a ranking for all reference and impaired sites sampled in that ecoregion or subecoregion.

Metrics with discrimination efficiencies (DE) above 0.80 were preferred for index development. In some cases, metrics with lower DE values were used in indices when "stronger" metrics were not available (as was the case particularly at the ecoregional level and in some of the coastal plain subecoregions). In no case should metrics with DE < .5 be used in any index. For further discussion of discrimination efficiencies and metric selection, please refer to Chapters 5 and 6.

ECOREGION 45—PIEDMONT

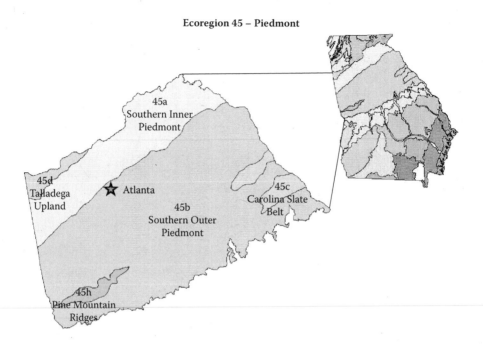

FIGURE F.1   Ecoregion 45—Piedmont.

Considered the nonmountainous portion of the old Appalachians Highland by physiographers, the northeast–southwest trending Piedmont ecoregion comprises a transitional area between the mostly mountainous ecoregions of the Appalachians to the northwest and the relatively flat coastal plain to the southeast. It is a complex mosaic of Precambrian and Paleozoic metamorphic and igneous rocks, with moderately dissected irregular plains and some hills. The soils tend to be finer textured than in coastal plain regions (Ecoregions 63 and 65). Once largely cultivated, much of this region has reverted to successional pine and hardwood woodlands, with an increasing conversion to an urban and suburban land cover.

TABLE F.1
Characteristic Reference Stream Land Use, Habitat, and Chemistry Data for Ecoregion 45—Piedmont

	Parameter	Mean	Median	Range
Catchment	% Natural	82.7	84.5	40.3–95.6
Land Use	% Agriculture	9.0	7.0	0–29.8
	% Silviculture	8.4	7.0	3.0–18.7
	% Urban	5.7	5.7	2.8–8.8
Habitat	Total Habitat Score (200)	155.6	160.0	128–184
	Epifaunal Substrate (20)	15.7	16.0	8–19
	Embeddedness or Pool Substrate Characterization (20)	14.4	15.0	5–18
	Velocity/Depth Regime or Pool Variability (20)	15.1	16.0	8–19
	Sediment Deposition (20)	13.5	14.0	6–19
	Channel Flow Status (20)	14.7	15.0	7–18
	Channel Alteration (20)	17.5	18.0	15–19
	Frequency of Riffles or Channel Sinuosity (20)	15.4	16.0	9–20
	Bank Stability (L) (10)	7.5	8.0	5–10
	Bank Stability (R) (10)	7.8	8.0	5–10
	Vegetative Protection (L) (10)	8.1	8.0	4–9
	Vegetative Protection (R) (10)	8.1	9.0	4–9
	Riparian Vegetative Width (L) (10)	9.0	9.0	7–10
	Riparian Vegetative Width (R) (10)	8.7	9.0	5–10
In-Stream	% Silt/Clay	3.7	1.0	0–42.0
Habitat	% Sand	36.2	26.2	2.8–89.0
(substrate)	% Gravel	34.9	35.0	2.0–78.0
	% Cobble	18.3	14.1	0–50.0
	% Boulder	4.2	2.0	0–33.0
	% Bedrock	2.7	0.0	0–20.6
Chemistry	Specific Conductivity (mS/cm)	0.113	0.052	0.03–1.21
(*in situ*)	Dissolved Oxygen (mg/l)	8.9	8.9	2.31–13.77
	pH (SU)	6.9	6.9	6.5–7.4
	Turbidity (NTU)	7.1	3.9	0–30.6
Chemistry	Alkalinity (mg/l as $CaCO_3$)	27.9	16.9	6.7–88.2
(laboratory)	Total Hardness (mg/l as $CaCO_3$)	26.1	20.1	6.7–87.2
	Ammonia (mg/l as N)	0.10	0.05	BD–0.97
	Nitrate–Nitrite (mg/l as N)	0.03	0.02	BD–0.08
	Total Phosphorous (mg/l as P)	0.31	0.17	BD–1.17
	Copper (mg/l)	0.007	0.007	BD–0.009
	Iron (mg/l)	2.33	1.12	BD–9.79
	Manganese (mg/l)	0.39	0.35	BD–0.77
	Zinc (mg/l)	0.03	0.02	BD–0.07

Note: BD = Below detection.

TABLE F.2
Discriminating Invertebrate Metrics for Ecoregion 45—Piedmont

Index 45

Metric	Metric Category
Coleoptera Taxa	Richness
% Chironomidae	Composition
% Plecoptera	
% Intolerant Individuals	Tolerance/Intolerance
North Carolina Biotic Index (NCBI)	

TABLE F.3
Descriptive Statistics for Reference Streams in Ecoregion 45—Piedmont

Metrics	DE	Minimum	Percentile (n = 23)					Maximum
			5th	25th	50th	75th	95th	
Coleoptera Taxa	0.6	0.0	1.1	4.5	5.0	8.0	10.8	12.0
% Chironomidae	0.7	3.0	7.4	19.6	33.2	41.6	58.2	59.6
% Plecoptera	0.7	0.4	0.4	3.4	6.9	15.5	38.9	74.1
% Intolerant Individuals	0.6	3.4	4.0	13.6	23.3	31.2	50.1	78.0
North Carolina Biotic Index (NCBI)	0.6	2.1	3.9	4.8	5.4	6.2	7.7	7.8

Note: n = Number of reference sites used; DE = discrimination efficiency between reference and impaired conditions.

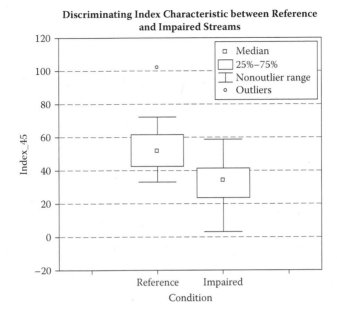

FIGURE F.2 Discriminating index characteristics between reference and impaired streams in Ecoregion 45—Piedmont.

TABLE F.4
Description of Numeric Rankings for Ecoregion 45—Piedmont

Index Score	Numeric Ranking	Percentile (n = 65)
67 and above	1	Above 95th
50–66	2	Below 95th, above 75th
31–49	3	Below 75th, above 25th
12–30	4	Below 25th, above 5th
11 and below	5	Below 5th

Note: n = All reference and impaired sites in Ecoregion 45.

TABLE F.5
Stream Ratings for Subecoregion 45—Piedmont

Station ID	Subecoregion	Candidate Condition	Index Score	Numeric Ranking	Narrative Description	Stream Rating
45d-15	45d	Reference	95	1	Very good	A
45a-89	45a	Reference	72	1	Very good	A
45h-13	45h	Reference	71	1	Very good	A
45a-3	45a	Reference	69	1	Very good	A
HH16	45a	Reference	62	2	Good	A
45b-152	45b	Reference	62	2	Good	A
45d-14	45d	Impaired	59	2	Good	A
45d-4	45d	Reference	57	2	Good	A
45d-16	45d	Reference	57	2	Good	A
45c-19	45c	Reference	56	2	Good	A
45h-1	45h	Impaired	56	2	Good	A
45a-90	45a	Impaired	54	2	Good	A
45d-9	45d	Reference	53	2	Good	A
45c-8	45c	Reference	53	2	Good	A
45h-9	45h	Reference	52	2	Good	A
45h-6	45h	Reference	50	2	Good	A
45d-21	45d	Impaired	50	2	Good	A
45d-11	45d	Impaired	50	2	Good	A
45c-16	45c	Reference	49	3	Fair	B
45b-258	45b	Reference	49	3	Fair	B
45a-38	45a	Impaired	47	3	Fair	B
45h-17	45h	Reference	47	3	Fair	B
45b-1214	45b	Impaired	46	3	Fair	B
45a-35	45a	Impaired	46	3	Fair	B
HH22	45b	Reference	44	3	Fair	B
45b-201	45b	Impaired	43	3	Fair	B
45b-357	45b	Reference	42	3	Fair	B
45h-12	45h	Impaired	42	3	Fair	B
45b-120	45b	Impaired	41	3	Fair	B
45d-6	45d	Impaired	41	3	Fair	B
45h-11	45h	Impaired	40	3	Fair	B
HH18	45a	Reference	40	3	Fair	B
45h-16	45h	Reference	40	3	Fair	B
45b-10	45b	Impaired	39	3	Fair	B
45d-23	45d	Impaired	39	3	Fair	B
45c-17	45c	Impaired	37	3	Fair	B
45a03//	45a	Reference	36	3	Fair	B
45b-116	45b	Impaired	35	3	Fair	B
45b-3	45b	Impaired	35	3	Fair	B
45h-2	45h	Impaired	34	3	Fair	B

TABLE F.5 (CONTINUED)
Stream Ratings for Subecoregion 45—Piedmont

Station ID	Subecoregion	Candidate Condition	Index Score	Numeric Ranking	Narrative Description	Stream Rating
45d-8	45d	Ref/ Removed	34	3	Fair	B
45a-31	45a	Impaired	34	3	Fair	B
45c-7	45c	Impaired	34	3	Fair	B
//4	45c	Reference	33	3	Fair	B
45b-156	45b	Reference	33	3	Fair	B
45b-1213	45b	Impaired	33	3	Fair	B
45b-44	45b	Impaired	32	3	Fair	B
45c-10	45c	Impaired	31	3	Fair	B
45b-213	45b	Impaired	30	4	Poor	C
45b-1	45b	Impaired	29	4	Poor	C
45c-3	45c	Impaired	29	4	Poor	C
45b-193	45b	Impaired	25	4	Poor	C
45c-11	45c	Impaired	25	4	Poor	C
45b-13	45b	Impaired	24	4	Poor	C
45b-212	45b	Impaired	23	4	Poor	C
45b-203	45b	Impaired	23	4	Poor	C
45a-61	45a	Impaired	22	4	Poor	C
45c-18	45c	Ref/ Removed	20	4	Poor	C
45h-10	45h	Impaired	19	4	Poor	C
45a-55	45a	Impaired	17	4	Poor	C
45b-202	45b	Impaired	12	4	Poor	C
45a-50	45a	Impaired	11	5	Very poor	C
45a-59	45a	Impaired	8	5	Very poor	C
45b-217	45b	Impaired	5	5	Very poor	C
45b-291	45b	Impaired	4	5	Very poor	C

SUBECOREGION 45B—SOUTHERN OUTER PIEDMONT

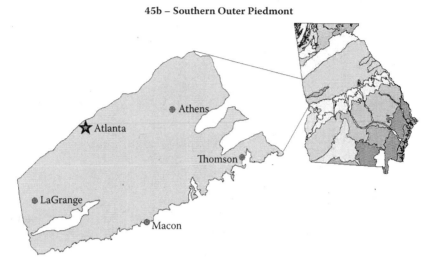

FIGURE F.3 Subecoregion 45b—Southern Outer Piedmont.

The Southern Outer Piedmont has lower elevations, less relief, and less precipitation than Subecoregion 45a. Loblolly-shortleaf pine is the major forest type, with less oak-hickory and oak-pine than in 45a. Gneiss, schist, and granite are the dominant rock types, covered with deep saprolite and mostly red, clayey subsoils. The majority of soils are Kanhapludults. The southern boundary of the subecoregion occurs at the Fall Line, where unconsolidated coastal plain sediments are deposited over the Piedmont metamorphic and igneous rocks.

TABLE F.6
Characteristic Reference Stream Land Use, Habitat, and Chemistry Data for Subecoregion 45b—Southern Outer Piedmont

	Parameter	Mean	Median	Range
Catchment	% Natural	86.7	86.8	80.2–92.9
Land Use	% Agriculture	0.90	0.50	0.1–2.3
	% Silviculture	7.0	5.2	3.0–14.7
	% Urban	5.4	5.6	4.0–6.5
Habitat	Total Habitat Score (200)	148.8	153.0	128–161
	Epifaunal Substrate (20)	14.6	15.0	8–19
	Embeddedness or Pool Substrate Characterization (20)	12.0	13.0	5–16
	Velocity/Depth Regime or Pool Variability (20)	15.6	16.0	13–17
	Sediment Deposition (20)	11.2	13.0	6–15
	Channel Flow Status (20)	15.8	17.0	13–18
	Channel Alteration (20)	17.0	17.0	15–19
	Frequency of Riffles or Channel Sinuosity (20)	14.6	15.0	12–16
	Bank Stability (L) (10)	6.8	7.0	6–8
	Bank Stability (R) (10)	7.0	7.0	5–9
	Vegetative Protection (L) (10)	8.0	8.0	7–9
	Vegetative Protection (R) (10)	8.4	8.5	8–9
	Riparian Vegetative Width (L) (10)	9.2	10.0	7–10
	Riparian Vegetative Width (R) (10)	8.6	9.0	5–10
In-Stream	% Silt/Clay	9.4	1.9	0–42.0
Habitat	% Sand	56.1	55.3	24.8–89.0
(substrate)	% Gravel	23.9	29.1	2.0–43.7
	% Cobble	6.5	1.0	0–22.8
	% Boulder	1.2	1.0	0–2.9
	% Bedrock	4.0	3.0	0–9.9
Chemistry	Specific Conductivity (mS/cm)	0.081	0.082	0.044–0.106
(*in situ*)	Dissolved Oxygen (mg/l)	9.7	7.8	7.2–13.8
	pH (SU)	6.9	6.8	6.6–7.2
	Turbidity (NTU)	10.0	10.5	0–20.3
Chemistry	Alkalinity (mg/l as $CaCO_3$)	24.9	32.5	0–44.9
(laboratory)	Total Hardness (mg/l as $CaCO_3$)	27.7	26.6	13.2–40.2
	Ammonia (mg/l as N)	0.36	0.07	BD–0.97
	Nitrate–Nitrite (mg/l as N)	0.08	0.09	0.01–0.16
	Total Phosphorous (mg/l as P)	0.042	0.042	BD–0.042
	Copper (mg/l)	0.004	0.004	BD–0.004
	Iron (mg/l)	1.12	1.12	BD–1.12
	Manganese (mg/l)	0.11	0.11	BD–0.11
	Zinc (mg/l)	0.01	0.01	BD–0.01

Note: BD = Below detection.

FIGURE F.4   Typical reference stream for Subecoregion 45b—Southern Outer Piedmont.

FIGURE F.5   Typical impaired stream for Subecoregion 45b—Southern Outer Piedmont.

TABLE F.7
Discriminating Invertebrate Metrics for Subecoregion 45b—Southern Outer Piedmont

Index 45b	
Metric	**Metric Category**
Coleoptera Taxa	Richness
% Oligochaeta	Composition
% Plecoptera	
Shredder Taxa	Functional Feeding Group
Scraper Taxa	
Swimmer Taxa	Habit

TABLE F.8
Descriptive Statistics for Reference Streams in Subecoregion 45b—Southern Outer Piedmont

Metrics	DE	Minimum	Percentile (*n* = 5)					Maximum
			5th	25th	50th	75th	95th	
Coleoptera Taxa	0.9	5.0	5.2	6.0	6.0	8.0	8.8	9.0
% Oligochaeta	0.8	0.0	0.0	0.0	0.8	1.2	2.9	3.0
% Plecoptera	0.8	1.0	2.0	4.6	10.8	12.9	16.2	17.0
Shredder Taxa	0.9	4.0	4.4	6.0	6.0	7.0	11.0	12.0
Scraper Taxa	0.9	2.0	2.4	4.0	6.0	8.0	8.8	9.0
Swimmer Taxa	0.9	1.0	1.4	3.0	3.0	3.0	3.8	4.0

Note: *n* = Number of reference sites used; DE = discrimination efficiency between reference and impaired conditions.

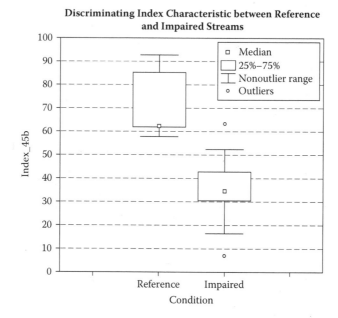

FIGURE F.6 Discriminating index characteristics between reference and impaired streams for Subecoregion 45b—Southern Outer Piedmont.

TABLE F.9

Description of Numeric Rankings for Subecoregion 45b—Southern Outer Piedmont

Index Score	Numeric Ranking	Percentile ($n = 22$)
84 and above	1	Above 95th
56–83	2	Below 95th, above 75th
32–55	3	Below 75th, above 25th
17–31	4	Below 25th, above 5th
16 and below	5	Below 5th

Note: n = All reference and impaired sites in Subecoregion 45b.

TABLE F.10
Stream Ratings for Subecoregion 45b—Southern Outer Piedmont

Condition	Index Score	Numeric Ranking	Narrative Description	Stream Rating
Reference	93	1	Very good	A
Reference	85	1	Very good	A
Impaired	63	2	Good	A
Reference	62	2	Good	A
Reference	62	2	Good	A
Reference	58	2	Good	A
Impaired	53	3	Fair	B
Impaired	51	3	Fair	B
Impaired	46	3	Fair	B
Impaired	43	3	Fair	B
Impaired	39	3	Fair	B
Impaired	38	3	Fair	B
Impaired	35	3	Fair	B
Impaired	34	3	Fair	B
Impaired	34	3	Fair	B
Impaired	31	4	Poor	B
Impaired	31	4	Poor	B
Impaired	30	4	Poor	B
Impaired	23	4	Poor	B
Impaired	20	4	Poor	B
Impaired	16	5	Very poor	B
Impaired	7	5	Very poor	B

ECOREGION 65—SOUTHEASTERN PLAINS

Ecoregion 65 – Southeastern Plains

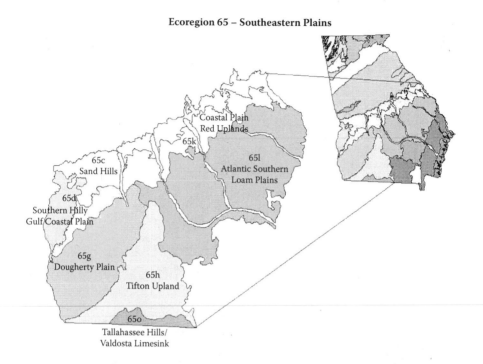

FIGURE F.7   Ecoregion 65—Southeastern Plains.

The Southeastern Plains are irregular plains with broad interstream areas that have a mosaic of cropland, pasture, woodland, and forest. Natural vegetation is mostly oak-hickory-pine and Southern mixed forest. The Cretaceous or Tertiary-age sands, silts, and clays of the region contrast geologically with the Paleozoic limestone, shale, and sandstone of Ecoregions 67 and 68, or with the even older metamorphic and igneous rocks of the Piedmont (Ecoregion 45). Elevations and relief are greater than in the Southern Coastal Plain (Ecoregion 75), but generally less than in much of the Piedmont. Streams in this area are relatively low gradient and sandy bottomed.

TABLE F.11
Characteristic Reference Stream Land Use, Habitat, and Chemistry Data for Ecoregion 65—Southeastern Plains

	Parameter	Mean	Median	Range
Catchment	% Natural	71.3	73.9	43.3–94.8
Land Use	% Agriculture	22.5	21.2	0.7–51.2
	% Silviculture	7.8	6.4	0.2–24.8
	% Urban	5.7	5.5	0.7–11.8
Habitat	Total Habitat Score (200)	158.2	161.5	121–179
	Epifaunal Substrate (20)	15.6	16.0	11–18
	Pool Substrate Characterization (20)	14.3	15.0	7–18
	Pool Variability (20)	14.7	16.0	7–19
	Sediment Deposition (20)	15.3	16.0	6–18
	Channel Flow Status (20)	16.5	17.0	10–19
	Channel Alteration (20)	17.2	17.0	15–20
	Channel Sinuosity (20)	14.0	14.5	5–20
	Bank Stability (L) (10)	8.3	9.0	4–10
	Bank Stability (R) (10)	8.3	9.0	4–10
	Vegetative Protection (L) (10)	8.9	9.0	7–10
	Vegetative Protection (R) (10)	9.1	9.0	6–10
	Riparian Vegetative Width (L) (10)	8.0	8.0	4–10
	Riparian Vegetative Width (R) (10)	8.0	8.0	4–10
In-Stream	% Silt/Clay	17.5	7.0	0.0–100.0
Habitat	% Sand	79.7	90.5	0.0–100.0
(substrate)	% Gravel	2.7	0.0	0.0–30.5
	% Cobble	0.6	0.0	0.0–16.7
	% Boulder	0.1	0.0	0.0–1.9
	% Bedrock	0.1	0.0	0.0–2.0
Chemistry	Specific Conductivity (mS/cm)	0.097	0.060	0.036–0.089
(*in situ*)	Dissolved Oxygen (mg/l)	9.3	9.3	5.5–16.5
	pH (SU)	6.1	6.3	4.1–7.5
	Turbidity (NTU)	8.3	6.9	0.0–39.6
Chemistry	Alkalinity (mg/l as $CaCO_3$)	23.2	8.6	0.0–176.0
(laboratory)	Total Hardness (mg/l as $CaCO_3$)	37.0	21.2	0.0–196.9
	Ammonia (mg/l as N)	0.056	0.054	BD–0.089
	Nitrate–Nitrite (mg/l as N)	0.160	0.076	BD–0.806
	Total Phosphorous (mg/l as P)	0.085	0.054	BD–0.209
	Copper (mg/l)	0.003	0.003	BD–0.003
	Iron (mg/l)	1.98	0.82	BD–12.99
	Manganese (mg/l)	0.092	0.092	BD–0.141
	Zinc (mg/l)	0.036	0.034	BD–0.052

Note: BD = Below detection.

TABLE F.12
Discriminating Invertebrate Metrics for Ecoregion 65—Southeastern Plains

Index 65	
Metric	**Metric Category**
% Coleoptera	Composition
% Oligochaeta	
Intolerant Taxa	Tolerance/Intolerance
% Intolerant Individuals	
% Predator	Functional Feeding Group
% Clinger	Habit

TABLE F.13
Descriptive Statistics for Reference Streams in Ecoregion 65—Southeastern Plains

Metrics	DE	Minimum	5th	25th	50th	75th	95th	Maximum
% Coleoptera	0.5	0.0	0.2	2.0	4.6	9.8	17.8	36.4
% Oligochaeta	0.6	0.0	0.0	0.0	1.4	2.9	10.4	11.7
Intolerant Taxa	0.5	0.0	0.0	2.0	3.5	6.0	9.9	12.0
% Intolerant Individuals	0.6	0.0	0.0	2.5	6.4	13.6	36.23	46.7
% Predator	0.5	1.5	3.1	7.5	11.8	19.3	38.5	48.8
% Clinger	0.5	0.0	2.5	8.1	18.8	27.6	47.8	63.3

The 5th through 95th columns fall under the header: **Percentile (n = 32)**

Note: n = Number of reference sites used; DE = discrimination efficiency between reference and impaired conditions.

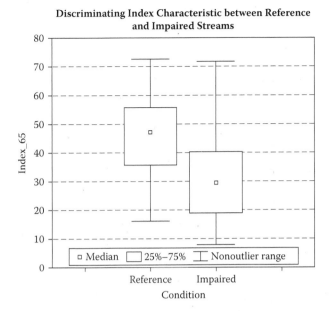

FIGURE F.8 Discriminating index characteristics between reference and impaired streams in Ecoregion 65—Southeastern Plains.

TABLE F.14
Description of Numeric Rankings for Ecoregion 65—Southeastern Plains

Index Score	Numeric Ranking	Percentile ($n = 103$)
63 and above	1	Above 95th
49–62	2	Below 95th, above 75th
23–48	3	Below 75th, above 25th
16–22	4	Below 25th, above 5th
15 and below	5	Below 5th

Note: n = All reference and impaired sites in Ecoregion 65.

TABLE F.15
Stream Ratings for Ecoregion 65—Southeastern Plains

Station ID	Subecoregion	Condition	Index Score	Numeric Ranking	Narrative Description	Stream Rating
HH25	65c	Reference	73	1	Very good	A
65c-40	65c	Impaired	72	1	Very good	A
65d-14	65d	Reference	71	1	Very good	A
65c-12	65c	Impaired	65	1	Very good	A
65k-55	65k	Reference	64	1	Very good	A
65d-39	65d	Impaired	62	2	Good	A
65h-212	65h	Reference	62	2	Good	A
65l-146	65l	Impaired	62	2	Good	A
65k-127	65k	Impaired	60	2	Good	A
HH29	65g	Reference	59	2	Good	A
65h-206	65h	Reference	58	2	Good	A
HH26	65c	Reference	58	2	Good	A
65o-23	65o	Reference	56	2	Good	A
65d-18	65d	Reference	56	2	Good	A
HH24	65c	Reference	55	2	Good	A
65o-3	65o	Impaired	55	2	Good	A
65c-8	65c	Impaired	54	2	Good	A
65c-89	65c	Reference	53	2	Good	A
65h-202	65h	Reference	53	2	Good	A
65d-4	65d	Reference	53	2	Good	A
65d-21	65d	Impaired	52	2	Good	A
65g-83	65g	Ref. Removed	52	2	Good	A
65d-17	65d	Impaired	51	2	Good	A
65c-80	65c	Reference	51	2	Good	A
65h-174	65h	Impaired	50	2	Good	A
65h-37	65h	Impaired	48	3	Fair	B
65o-24	65o	Reference	48	3	Fair	B
65d-20	65d	Impaired	47	3	Fair	B
65g-62	65g	Reference	47	3	Fair	B
65l-10	65l	Reference	47	3	Fair	B
65h-203	65h	Reference	47	3	Fair	B
65k-115	65k	Impaired	46	3	Fair	B
65k-102	65k	Impaired	46	3	Fair	B
65k-100	65k	Impaired	46	3	Fair	B
65h-14	65h	Impaired	45	3	Fair	B
65o-12	65o	Reference	45	3	Fair	B
65d-3	65d	Reference	44	3	Fair	B
65k-54	65k	Reference	43	3	Fair	B
65k-56	65k	Reference	43	3	Fair	B

TABLE F.15 (CONTINUED)
Stream Ratings for Ecoregion 65—Southeastern Plains

Station ID	Subecoregion	Condition	Index Score	Numeric Ranking	Narrative Description	Stream Rating
65d-38	65d	Reference	42	3	Fair	B
65c-5	65c	Impaired	41	3	Fair	B
65c-3	65c	Impaired	40	3	Fair	B
65h-41	65h	Impaired	40	3	Fair	B
65k-129	65k	Impaired	40	3	Fair	B
65k-128	65k	Impaired	40	3	Fair	B
65g-69	65g	Impaired	38	3	Fair	B
65c-88	65c	Impaired	38	3	Fair	B
65d-32	65d	Impaired	36	3	Fair	B
65g-17	65g	Impaired	36	3	Fair	B
65l-381	65l	Reference	36	3	Fair	B
65o-25	65o	Reference	35	3	Fair	B
65o-18	65o	Impaired	35	3	Fair	B
65k-37	65k	Impaired	34	3	Fair	B
65l-234	65l	Impaired	34	3	Fair	B
65k-68	65k	Reference	34	3	Fair	B
65g-120	65g	Reference	32	3	Fair	B
65c-38	65c	Impaired	32	3	Fair	B
65c-4	65c	Impaired	32	3	Fair	B
65L-184	65l	Impaired	32	3	Fair	B
65h-209	65h	Reference	32	3	Fair	B
65c-92	65c	Impaired	32	3	Fair	B
65c-48	65c	Impaired	31	3	Fair	B
65o-22	65o	Impaired	30	3	Fair	B
65h-24	65h	Impaired	30	3	Fair	B
65h-17	65h	Impaired	29	3	Fair	B
65l-342	65l	Reference	28	3	Fair	B
65o-11	65o	Impaired	28	3	Fair	B
65l-160	65l	Impaired	28	3	Fair	B
65l-420	65l	Impaired	27	3	Fair	B
65k-113	65k	Impaired	26	3	Fair	B
65d-1	65d	Impaired	25	3	Fair	B
65l-343	65l	Reference	25	3	Fair	B
65d-22	65d	Impaired	25	3	Fair	B
65k-18	65k	Impaired	25	3	Fair	B
65g-130	65g	Impaired	23	3	Fair	B
65k-110	65k	Impaired	23	3	Fair	B
65h-1	65h	Impaired	23	3	Fair	B
65g-10	65g	Impaired	22	4	Poor	C
65l-423	65l	Impaired	22	4	Poor	C

(Continued)

TABLE F.15 (CONTINUED)
Stream Ratings for Ecoregion 65—Southeastern Plains

Station ID	Subecoregion	Condition	Index Score	Numeric Ranking	Narrative Description	Stream Rating
65l-281	65l	Impaired	21	4	Poor	C
65k-99	65k	Impaired	21	4	Poor	C
65g-14	65g	Impaired	20	4	Poor	C
65l-379	65l	Reference	19	4	Poor	C
65l-391	65l	Impaired	19	4	Poor	C
65l-280	65l	Impaired	19	4	Poor	C
65l-235	65l	Impaired	19	4	Poor	C
65l-283	65l	Impaired	18	4	Poor	C
65h-4	65h	Impaired	18	4	Poor	C
65l-390	65l	Impaired	18	4	Poor	C
65g-135	65g	Impaired	17	4	Poor	C
65g-137	65g	Impaired	17	4	Poor	C
65k-85	65k	Reference	16	4	Poor	C
65h-34	65h	Impaired	16	4	Poor	C
65h-13	65h	Impaired	16	4	Poor	C
65h-5	65h	Impaired	16	4	Poor	C
65g-4	65g	Impaired	16	4	Poor	C
65g-82	65g	Ref. Removed	15	5	Very poor	C
65l-403	65l	Impaired	15	5	Very poor	C
65g-8	65g	Impaired	15	5	Very poor	C
65h-32	65h	Impaired	14	5	Very poor	C
65l-277	65l	Impaired	12	5	Very poor	C
65g-84	65g	Impaired	11	5	Very poor	C
65o-9	65o	Impaired	8	5	Very poor	C

SUBECOREGION 65C—SAND HILLS

65c – Sand Hills

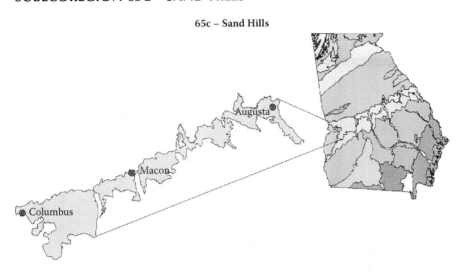

FIGURE F.9   Subecoregion 65c—Sand Hills.

The Sand Hills of Georgia form a narrow, rolling to hilly, highly dissected coastal plain belt stretching across the state from Augusta to Columbus. The region is composed primarily of Cretaceous and some Eocene-age marine sands and clays deposited over the crystalline and metamorphic rocks of the Piedmont (Ecoregion 45). Many of the droughty, low-nutrient soils formed in thick beds of sand, although soils in some areas contain more loamy and clayey horizons. On the drier sites, turkey oak and longleaf pine are dominant, while shortleaf-loblolly pine forests and other oak-pine forests are common throughout the region.

TABLE F.16
Characteristic Reference Stream Land Use, Habitat, and Chemistry Data for Subecoregion 65c–Sand Hills

	Parameter	Mean	Median	Range
Catchment	% Natural	72.5	72.2	65.4–77.5
Land Use	% Agriculture	7.1	8.4	0–13.1
	% Silviculture	15.3	15.3	9.0–21.1
	% Urban	5.1	5.2	3.0–7.3
Habitat	Total Habitat Score (200)	164.4	164.0	159–170
	Epifaunal Substrate (20)	15.6	16.0	13–18
	Pool Substrate Characterization (20)	13.8	15.0	9–16
	Pool Variability (20)	14.8	16.0	10–16
	Sediment Deposition (20)	17.0	17.0	16–18
	Channel Flow Status (20)	19.0	19.0	19
	Channel Alteration (20)	18.4	19.0	17–19
	Channel Sinuosity (20)	11.8	13.0	9–15
	Bank Stability (L) (10)	8.8	9.0	8–9
	Bank Stability (R) (10)	9.2	9.0	8–10
	Vegetative Protection (L) (10)	8.4	8.0	8–9
	Vegetative Protection (R) (10)	8.4	8.0	8–9
	Riparian Vegetative Width (L) (10)	9.4	10.0	8–10
	Riparian Vegetative Width (R) (10)	9.8	10.0	9–10
In-Stream	% Silt/Clay	37.0	12.0	0–22.8
Habitat	% Sand	87.0	95.7	63.0–100.0
(substrate)	% Gravel	1.1	0	0–4.3
	% Cobble	0	0	0
	% Boulder	0	0	0
	% Bedrock	0	0	0
Chemistry	Specific Conductivity (mS/cm)	0.020	0.015	0.003–0.049
(*in situ*)	Dissolved Oxygen (mg/l)	11.3	11.7	10.3–12.5
	pH (SU)	5.1	5.1	4.3–6.2
	Turbidity (NTU)	2.3	1.1	0–6.9
Chemistry	Alkalinity (mg/l as $CaCO_3$)	1.8	0	0–8.2
(laboratory)	Total Hardness (mg/l as $CaCO_3$)	9.8	10.3	5.5–18.0
	Ammonia (mg/l as N)	0.054	0.052	BD–0.07
	Nitrate–Nitrite (mg/l as N)	0.18	0.11	0.07–0.47
	Total Phosphorous (mg/l as P)	BD	BD	BD
	Copper (mg/l)	BD	BD	BD
	Iron (mg/l)	0.54	0.54	BD–0.92
	Manganese (mg/l)	BD	BD	BD
	Zinc (mg/l)	BD	BD	BD

Note: BD = Below detection.

FIGURE F.10   Typical reference stream for Subecoregion 65c—Sand Hills.

FIGURE F.11   Typical impaired stream for Subecoregion 65c—Sand Hills.

TABLE F.17
Discriminating Invertebrate Metrics for Subecoregion
65c—Sand Hills

Index 65c	
Metric	Metric Category
% Trichoptera	Composition
Tolerant Taxa	Tolerance/Intolerance
Intolerant Taxa	
% Scraper	Functional Feeding Group
Clinger Taxa	Habit

TABLE F.18
Descriptive Statistics for Reference Streams in Subecoregion 65c—Sand Hills

Metrics	DE	Minimum	Percentile (*n* = 5)					Maximum
			5th	25th	50th	75th	95th	
% Trichoptera	0.7	4.3	4.5	5.1	8.8	13.7	23.8	26.3
Tolerant Taxa	0.8	3.0	3.8	7.0	10.0	11.0	11.8	12.0
Intolerant Taxa	0.8	3.0	3.4	5.0	5.0	9.0	10.6	11.0
% Scraper	0.9	4.0	5.0	10.8	11.3	23.6	27.1	28.0
Clinger Taxa	0.6	10.0	10.2	11.0	12.0	15.0	16.6	17.0

Note: *n* = Number of reference sites used; DE = discrimination efficiency between reference and impaired conditions.

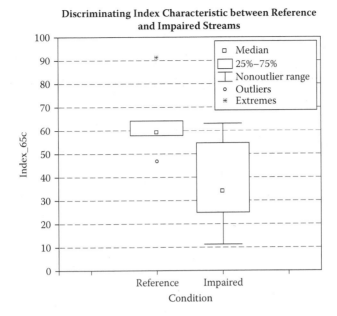

FIGURE F.12 Discriminating index characteristics between reference and impaired streams for Subecoregion 65c—Sand Hills.

TABLE F.19
Description of Numeric Rankings for Subecoregion 65c—Sand Hills

Index Score	Numeric Ranking	Percentile ($n = 15$)
73 and above	1	Above 95th
61–72	2	Below 95th, above 75th
30–60	3	Below 75th, above 25th
20–29	4	Below 25th, above 5th
19 and below	5	Below 5th

Note: n = All reference and impaired sites in Subecoregion 65c.

TABLE F.20
Stream Ratings for Subecoregion 65c—Sand Hills

Station ID	Condition	Index Score	Numeric Ranking	Narrative Description	Stream Rating
HH25	Reference	92	1	Very good	A
HH24	Reference	65	2	Good	A
65c-40	Impaired	63	2	Good	A
65c-3	Impaired	62	2	Good	A
65c-80	Reference	59	3	Fair	B
65c-89	Reference	58	3	Fair	B
65c-8	Impaired	55	3	Fair	B
65c-12	Impaired	52	3	Fair	B
HH26	Reference	47	3	Fair	B
65c-88	Impaired	35	3	Fair	B
65c-5	Impaired	34	3	Fair	B
65c-38	Impaired	26	4	Poor	C
65c-92	Impaired	25	4	Poor	C
65c-48	Impaired	24	4	Poor	C
65c-4	Impaired	11	5	Very poor	C

ECOREGION 66—BLUE RIDGE

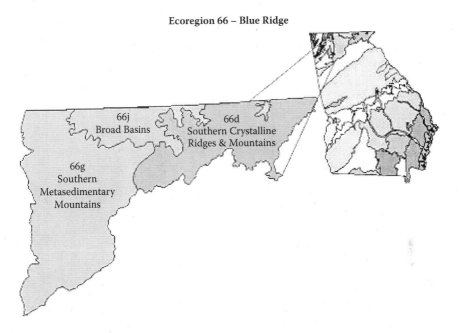

FIGURE F.13   Ecoregion 66—Blue Ridge.

The Blue Ridge extends from southern Pennsylvania to northern Georgia, varying from narrow ridges to hilly plateaus to more massive mountainous areas with high peaks. The mostly forested slopes; high-gradient, cool, clear streams; and rugged terrain occur on a mix of igneous, metamorphic, and sedimentary geology. Annual precipitation of over 80 inches can occur on the well-exposed high peaks. The southern Blue Ridge is one of the richest centers of biodiversity in the eastern United States. It is one of the most floristically diverse ecoregions, and includes Appalachian oak forests, northern hardwoods, and, at the highest elevations in Tennessee and North Carolina, Southeastern spruce-fir forests. Shrub, grass, heath balds, hemlock, cove hardwoods, and oak-pine communities are also significant. Black bear, whitetail deer, wild boar, turkey, grouse, songbirds, many species of amphibians and reptiles, thousands of species of invertebrates, and a variety of small mammals are found here.

TABLE F.21
Characteristic Reference Stream Land Use, Habitat, and Chemistry Data for Ecoregion 66—Blue Ridge

	Parameter	Mean	Median	Range
Catchment	% Natural	91.3	95.9	61.6–99.8
Land Use	% Agriculture	2.9	0.2	0.0–11.7
	% Silviculture	0.4	0.1	0.0–4.0
	% Urban	2.4	1.9	0.0–6.4
Habitat	Total Habitat Score (200)	166.7	167.5	111–192
	Epifaunal Substrate (20)	16.1	16.5	10–19
	Embeddedness (20)	16.0	17.5	4–19
	Velocity/Depth Regime (20)	16.6	17.0	14–19
	Sediment Deposition (20)	15.6	17.0	5–19
	Channel Flow Status (20)	16.5	16.5	13–20
	Channel Alteration (20)	18.1	18.0	15–20
	Frequency of Riffles (20)	17.9	18.0	16–20
	Bank Stability (L) (10)	8.8	9.0	4–10
	Bank Stability (R) (10)	8.5	9.0	3–10
	Vegetative Protection (L) (10)	8.4	9.0	3–10
	Vegetative Protection (R) (10)	7.9	9.0	3–10
	Riparian Vegetative Width (L) (10)	8.9	10.0	1–10
	Riparian Vegetative Width (R) (10)	7.6	8.5	1–10
In-Stream Habitat	% Silt/Clay	2.6	0.0	0.0–12.1
(substrate)	% Sand	8.9	7.0	0.0–28.0
	% Gravel	37.5	35.0	13.2–69.0
	% Cobble	34.3	32.4	4.0–54.0
	% Boulder	14.8	14.0	0.0–33.0
	% Bedrock	1.9	0.0	0.0–8.0
Chemistry (in	Specific Conductivity (mS/cm)	0.017	0.016	0.008–0.038
situ)	Dissolved Oxygen (mg/l)	11.0	10.9	8.9–13.0
	pH (SU)	6.8	6.8	6.4–7.2
	Turbidity (NTU)	5.0	4.7	0.0–17.8
Chemistry	Alkalinity (mg/l as $CaCO_3$)	6.1	6.1	0.0–12.3
(laboratory)	Total Hardness (mg/l as $CaCO_3$)	6.6	6.5	2.7–15.4
	Ammonia (mg/l as N)	0.049	0.046	BD–0.036
	Nitrate–Nitrite (mg/l as N)	0.186	0.085	BD–0.841
	Total Phosphorous (mg/l as P)	0.089	0.075	BD–0.062
	Copper (mg/l)	BD	BD	BD
	Iron (mg/l)	0.151	0.102	BD–0.458
	Manganese (mg/l)	0.010	0.006	BD–0.029
	Zinc (mg/l)	0.011	0.012	BD–0.006

Note: BD = Below detection.

TABLE F.22
Discriminating Invertebrate Metrics for the Ecoregion 66—Blue Ridge

Index 66

Metric	Metric Category
Plecoptera Taxa	Richness
Simpson's Index	
% Trichoptera	Composition
% Intolerant Individuals	
NCBI	Tolerance/Intolerance
Predator Taxa	Functional Feeding Group
Burrower Taxa	Habit

TABLE F.23
Descriptive Statistics for Reference Streams in Ecoregion 66—Blue Ridge

Metrics	DE	Minimum	Percentile (n = 15)					Maximum
			5th	25th	50th	75th	95th	
Plecoptera Taxa	0.7	3.0	3.0	5.5	9.0	10.0	11.3	12.0
Simpson's Index	0.6	0.0	0.0	0.0	0.0	0.0	0.1	0.1
% Trichoptera	0.5	9.6	13.4	17.5	20.0	21.1	26.3	26.3
% Intolerant Individuals	0.6	12.5	18.5	24.6	37.9	48.3	52.0	54.7
NCBI	0.7	3.3	3.4	3.6	4.2	4.6	5.2	5.5
Predator Taxa	0.8	8.0	8.0	11.5	13.0	16.0	17.6	19.0
Burrower Taxa	0.5	4.0	4.7	6.0	7.0	8.5	9.6	11.0

Note: n = Number of reference sites used; DE = discrimination efficiency between reference and impaired conditions.

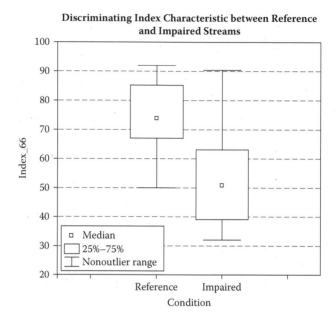

FIGURE F.14 Discriminating index characteristics between reference and impaired streams in Ecoregion 66—Blue Ridge.

TABLE F.24
Description of Numeric Rankings for Ecoregion 66—Blue Ridge

Index Score	Numeric Ranking	Percentile (*n* = 32)
90 and above	1	Above 95th
76–89	2	Below 95th, above 75th
49–75	3	Below 75th, above 25th
34–48	4	Below 25th, above 5th
33 and below	5	Below 5th

Note: *n* = All reference and impaired sites in Ecoregion 66.

TABLE F.25
Stream Ratings for Ecoregion 66—Blue Ridge

Station ID	Subecoregion	Condition	Index Score	Numeric Ranking	Narrative Description	Stream Rating
66g-23	66g	Reference	92	1	Very good	A
66d-48	66d	Impaired	90	1	Very good	A
66d-40	66d	Reference	90	1	Very good	A
66d-41	66d	Reference	88	2	Good	A
66g-5	66g	Reference	85	2	Good	A
66g-2-2	66g	Reference	80	2	Good	A
66d-38	66d	Impaired	80	2	Good	A
66j-23	66j	Reference	77	2	Good	A
66d-44-2	66d	Reference	75	3	Fair	B
66j-28	66j	Reference	74	3	Fair	B
66g-42	66g	Impaired	74	3	Fair	B
66j-211	66j	Reference	72	3	Fair	B
66d-44	66d	Reference	70	3	Fair	B
66j-26	66j	Impaired	68	3	Fair	B
66j-19	66j	Reference	67	3	Fair	B
66g-6	66g	Reference	67	3	Fair	B
66j-27	66j	Impaired	63	3	Fair	B
66j-31	66j	Reference	58	3	Fair	B
66g-71	66g	Impaired	56	3	Fair	B
66d-49	66d	Impaired	55	3	Fair	B
66d-58	66d	Reference	55	3	Fair	B
66j-9	66j	Impaired	53	3	Fair	B
66d-43	66d	Impaired	51	3	Fair	B
66g-2	66g	Reference	50	3	Fair	B
66g-39	66g	Impaired	43	4	Poor	C
66g-65	66g	Impaired	41	4	Poor	C
66d-50	66d	Impaired	39	4	Poor	C
66g-31	66g	Impaired	39	4	Poor	C
66j-25	66j	Impaired	39	4	Poor	C
66j-17	66j	Impaired	33	5	Very poor	C
66g-30	66g	Impaired	33	5	Very poor	C
66g-44	66g	Impaired	32	5	Very poor	C

66D—SOUTHERN CRYSTALLINE RIDGES AND MOUNTAINS

66d – Southern Crystalline Ridges and Mountains

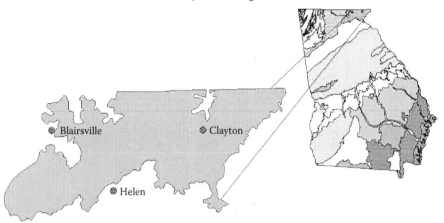

FIGURE F.15   66d—Southern Crystalline Ridges and Mountains.

The Southern Crystalline Ridges and Mountains contain the highest and wettest mountains in Georgia. These occur primarily on Precambrian-age igneous and high-grade metamorphic rocks. The common crystalline rock types include gneiss, schist, and quartzite, covered by well-drained, acidic, brownish, loamy soils. Some mafic and ultramafic rocks also occur here, producing more basic soils. Elevations of this rough, dissected region are typically 1800 to 4000 feet, with Brasstown Bald Mountain, the highest point in Georgia, reaching 4784 feet. Although there are a few small areas of pasture and apple orchards, the region is mostly forested.

TABLE F.26
Characteristic Reference Stream Land Use, Habitat, and Chemistry Data for Subecoregion 66d—Southern Crystalline Ridges and Mountains

	Parameter	Mean	Median	Range
Catchment	% Natural	97.7	97.9	97.2–98.1
Land Use	% Agriculture	0	0	0
	% Silviculture	0.30	0.30	0.10–0.50
	% Urban	1.9	1.9	1.5–2.6
Habitat	Total Habitat Score (200)	174.6	179.0	162–186
	Epifaunal Substrate (20)	16.8	17.0	14–19
	Embeddedness (20)	16.6	17.0	15–18
	Velocity/Depth Regime (20)	17.6	18.0	15–19
	Sediment Deposition (20)	16.0	16.0	13–18
	Channel Flow Status (20)	16.8	17.0	15–18
	Channel Alteration (20)	18.2	18.0	17–20
	Frequency of Riffles (20)	18.2	19.0	16–19
	Bank Stability (L) (10)	9.2	9.0	8–10
	Bank Stability (R) (10)	9.0	9.0	8–10
	Vegetative Protection (L) (10)	9.2	9.0	8–10
	Vegetative Protection (R) (10)	8.4	9.0	7–9
	Riparian Vegetative Width (L) (10)	9.8	10.0	9–10
	Riparian Vegetative Width (R) (10)	8.8	9.0	7–10
In-Stream	% Silt/Clay	0.2	0	0–1.0
Habitat	% Sand	6.2	6.0	2.9–8.0
(substrate)	% Gravel	33.1	32.0	28.4–40.2
	% Cobble	39.3	42.0	28.4–54.0
	% Boulder	17.4	17.6	6.0–30.0
	% Bedrock	3.8	3.0	0–8.0
Chemistry	Specific Conductivity (mS/cm)	0.013	0.012	0.008–0.016
(in situ)	Dissolved Oxygen (mg/l)	11.2	11.6	9.7–11.9
	pH (SU)	6.7	6.6	6.4–7.1
	Turbidity (NTU)	1.42	0.5	0–5.8
Chemistry	Alkalinity (mg/l as $CaCO_3$)	5.6	5.5	2.5–8.3
(laboratory)	Total Hardness (mg/l as $CaCO_3$)	6.3	4.0	3.5–10.4
	Ammonia (mg/l as N)	0.05	0.051	0.037–0.057
	Nitrate–Nitrite (mg/l as N)	0.052	0.063	BD–0.07
	Total Phosphorous (mg/l as P)	0.142	0.142	BD–0.142
	Copper (mg/l)	BD	BD	BD
	Iron (mg/l)	0.04	0.04	BD–0.04
	Manganese (mg/l)	0.006	0.006	BD–0.006
	Zinc (mg/l)	0.006	0.006	BD–0.006

Note: BD = Below detection.

FIGURE F.16 Typical reference stream for Subecoregion 66d—Southern Crystalline Ridge and Mountains.

FIGURE F.17 Typical impaired stream for Subecoregion 66d—Southern Crystalline Ridge and Mountains.

TABLE F.27
**Discriminating Invertebrate Metrics Subecoregion
66d—Southern Crystalline Ridge and Mountains**

Index 66d	
Metric	**Metric Category**
Diptera Taxa	Richness
% Plecoptera	Composition
% Odonata	
% Dominant Individuals	Tolerance/Intolerance
% Shredder	Functional Feeding Group
Clinger Taxa	Habit

TABLE F.28
**Descriptive Statistics for Reference Streams in Subecoregion 66d—Southern
Crystalline Ridges and Mountains**

Metrics	DE	Minimum	Percentile (*n* = 5)					Maximum
			5th	**25th**	**50th**	**75th**	**95th**	
Diptera Taxa	0.8	16.0	17.4	23.0	25.0	26.0	30.0	31.0
% Plecoptera	0.6	11.3	12.1	15.4	24.7	30.4	30.8	30.8
% Odonata	1.0	0.0	0.0	0.0	0.4	0.8	3.8	4.5
% Dominant Individuals	0.6	7.9	8.2	9.1	11.3	12.5	15.2	15.8
% Shredder	0.8	8.3	9.5	14.1	14.2	32.1	33.5	33.9
Clinger Taxa	0.6	15.0	15.8	19.0	22.0	23.0	30.2	32.0

Note: *n* = Number of reference sites used; DE = discrimination efficiency between reference and impaired conditions.

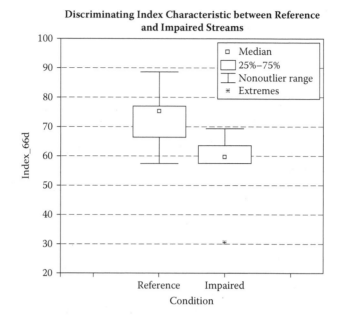

FIGURE F.18   Discriminating index characteristics between reference and impaired streams in Ecoregion 66d—Southern Crystalline Ridges and Mountains.

TABLE F.29
Description of Numeric Rankings for Subecoregion 66d—Southern Crystalline Ridges and Mountains

Index Score	Numeric Ranking	Percentile ($n = 10$)
83 and above	1	Above 95th
74–82	2	Below 95th, above 75th
58–73	3	Below 75th, above 25th
43–57	4	Below 25th, above 5th
42 and below	5	Below 5th

Note: n = All reference and impaired sites in Subecoregion 66d.

TABLE F.30
Stream Ratings for Subecoregion 66d—Southern Crystalline Ridges and Mountains

Station ID	Condition	Index Score	Numeric Ranking	Narrative Description	Stream Rating
66d-44	Reference	89	1	Very good	A
66d-44-2	Reference	77	2	Good	A
66d-40	Reference	75	2	Good	A
66d-43	Impaired	69	3	Fair	B
66d-58	Reference	66	3	Fair	B
66d-48	Impaired	64	3	Fair	B
66d-49	Impaired	60	3	Fair	B
66d-41	Reference	58	3	Fair	B
66d-38	Impaired	57	4	Poor	C
66d-50	Impaired	30	5	Very poor	C

ECOREGION 75—SOUTHERN COASTAL PLAIN

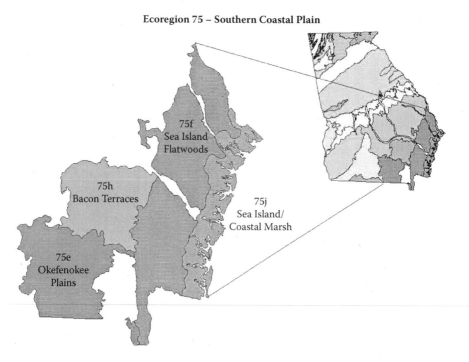

Ecoregion 75 – Southern Coastal Plain

FIGURE F.19   Ecoregion 75—Southern Coastal Plain.

The Southern Coastal Plain extends from South Carolina and Georgia through much of central Florida, and along the Gulf coast lowlands of the Florida Panhandle, Alabama, and Mississippi. From a national perspective, it appears to be mostly flat plains, but it is a heterogeneous region also containing barrier islands, coastal lagoons, marshes, and swampy lowlands along the Gulf and Atlantic coasts. In Florida, an area of discontinuous highlands contains numerous lakes. This ecoregion is generally lower in elevation with less relief and wetter soils than Ecoregion 65. Once covered by a variety of forest communities that included trees of longleaf pine, slash pine, pond pine, beech, sweetgum, southern magnolia, white oak, and laurel oak, land cover in the region is now mostly slash and loblolly pine with oak-gum-cypress forest in some low-lying areas, citrus groves, pasture for beef cattle, and urban.

TABLE F.31
Characteristic Reference Stream Land Use, Habitat, and Chemistry Data for Ecoregion 75—Southern Coastal Plain

	Parameter	Mean	Median	Range
Catchment	% Natural	87.0	90.5	64.5–95.9
Land Use	% Agriculture	4.9	1.5	0.0–27.1
	% Silviculture	11.8	11.3	0.2–35.0
	% Urban	8.0	6.5	4.1–20.5
Habitat	Total Habitat Score (200)	152.2	152.0	112–181
	Epifaunal Substrate (20)	14.8	15.0	10–19
	Pool Substrate Characterization (20)	13.0	13.0	8–19
	Pool Variability (20)	11.7	11.0	5–19
	Sediment Deposition (20)	15.8	17.5	8–20
	Channel Flow Status (20)	15.8	18.0	5–20
	Channel Alteration (20)	18.2	18.0	13–20
	Channel Sinuosity (20)	13.6	13.0	8–20
	Bank Stability (L) (10)	8.3	9.0	1–10
	Bank Stability (R) (10)	8.3	9.0	1–10
	Vegetative Protection (L) (10)	8.0	8.0	3–10
	Vegetative Protection (R) (10)	8.1	8.5	3–10
	Riparian Vegetative Width (L) (10)	8.6	9.0	5–10
	Riparian Vegetative Width (R) (10)	8.0	9.0	3–10
In-Stream	% Silt/Clay	28.2	12.9	0.0–100.0
Habitat	% Sand	71.4	85.6	0.0–100.0
(substrate)	% Gravel	0.4	0.0	0.0–4.0
	% Cobble	0.0	0.0	0.0
	% Boulder	0.0	0.0	0.0
	% Bedrock	0.0	0.0	0.0
Chemistry	Specific Conductivity (mS/cm)	0.871	0.108	0.051–8.920
(in situ)	Dissolved Oxygen (mg/l)	6.7	6.6	3.5–14.6
	pH (SU)	4.8	4.5	3.6–6.7
	Turbidity (NTU)	11.5	6.7	0.0–57.0
Chemistry	Alkalinity (mg/l as $CaCO_3$)	8.8	0.0	0.0–101.4
(laboratory)	Total Hardness (mg/l as $CaCO_3$)	135.5	33.2	7.7–1067.0
	Ammonia (mg/l as N)	5,397	0.083	BD–48.917
	Nitrate–Nitrite (mg/l as N)	0.117	0.051	BD–0.325
	Total Phosphorous (mg/l as P)	0.138	0.122	BD–0.323
	Copper (mg/l)	0.009	0.009	BD–0.015
	Iron (mg/l)	1.076	1.015	BD–2.897
	Manganese (mg/l)	0.040	0.036	BD–0.099
	Zinc (mg/l)	0.018	0.017	BD–0.023

Note: BD = Below detection.

TABLE F.32
Discriminating Invertebrate Metrics for Ecoregion 75—Southern Coastal Plain

Index 75	
Metric	**Metric Category**
% Noninsect	Composition
% Oligochaeta	
% Odonata	
% Tanypodinae/Total Chironomidae	
Tolerant Taxa	Tolerance/Intolerance
% Tolerant Individuals	

TABLE F.33
Descriptive Statistics for Reference Streams in Ecoregion 75—Southern Coastal Plain

Metrics	DE	Minimum	Percentile (*n* = 24)					Maximum
			5th	**25th**	**50th**	**75th**	**95th**	
% Noninsect	0.6	0.5	1.7	6.2	16.7	31.9	89.3	92.4
% Oligochaeta	0.7	0.0	0.0	0.7	1.0	3.9	6.5	8.1
% Odonata	0.5	0.0	0.0	0.0	0.0	0.5	3.3	9.2
% Tanypodinae/ Total Chironomidae	0.5	0.0	0.0	0.0	0.2	1.8	16.3	34.4
Tolerant Taxa	0.6	0.0	1.0	2.8	6.0	8.0	19.9	21.0
% Tolerant Individuals	0.5	0.0	1.0	11.7	29.8	53.8	68.8	93.3

Note: *n* = Number of reference sites used; DE = discrimination efficiency between reference and impaired conditions.

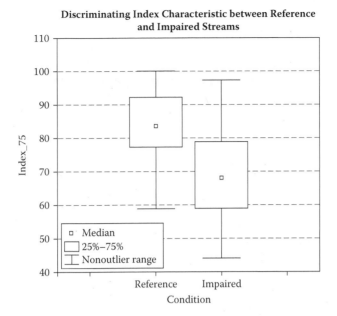

FIGURE F.20   Discriminating index characteristics between reference and impaired streams in Ecoregion 75—Southern Coastal Plain.

TABLE F.34
Description of Numeric Rankings for Ecoregion 75—Southern Coastal Plain

Index Score	Numeric Ranking	Percentile ($n = 60$)
94 and above	1	Above 95th
84–93	2	Below 95th, above 75th
65–83	3	Below 75th, above 25th
50–64	4	Below 25th, above 5th
49 and below	5	Below 5th

Note: n = All reference and impaired sites in Ecoregion 75.

TABLE F.35
Stream Ratings for Ecoregion 75—Southern Coastal Plain

Station ID	Subecoregion	Condition	Index Score	Numeric Ranking	Narrative Description	Stream Rating
75j-16	75j	Reference	100	1	Very good	A
75j-10	75j	Reference	98	1	Very good	A
75j-13	75j	Impaired	97	1	Very good	A
75h-35	75h	Reference	94	1	Very good	A
75f-91	75f	Reference	94	1	Very good	A
75j-15	75j	Reference	92	2	Good	A
75e-23	75e	Reference	92	2	Good	A
75h-45	75h	Reference	92	2	Good	A
75e-69	75e	Reference	91	2	Good	A
75h-10	75h	Reference	91	2	Good	A
75f-28	75f	Impaired	87	2	Good	A
75j-2	75j	Impaired	87	2	Good	A
75e-78	75e	Reference	87	2	Good	A
75j-26	75j	Reference	86	2	Good	A
75j-37	75j	Reference	85	2	Good	A
75j-21	75j	Impaired	84	2	Good	A
75j-24	75j	Impaired	84	2	Good	A
75j-4	75j	Impaired	83	3	Fair	B
75f-95	75f	Reference	82	3	Fair	B
75j-25	75j	Reference	82	3	Fair	B
75j-5	75j	Reference	81	3	Fair	B
75j-41	75j	Reference	81	3	Fair	B
75f-132	75f	Impaired	81	3	Fair	B
75f-124	75f	Reference	80	3	Fair	B
75f-48	75f	Impaired	80	3	Fair	B
75e-20	75e	Impaired	79	3	Fair	B
75j-31	75j	Reference	79	3	Fair	B
75f-127	75f	Impaired	78	3	Fair	B
75e-59	75e	Reference	76	3	Fair	B
75f-137	75f	Impaired	73	3	Fair	B
75f-70	75f	Impaired	72	3	Fair	B
75f-126	75f	Reference	71	3	Fair	B
75f-61	75f	Reference	70	3	Fair	B
75e-54	75e	Impaired	70	3	Fair	B
75h-47	75h	Impaired	70	3	Fair	B
75f-44	75f	Impaired	70	3	Fair	B
75j-23	75j	Impaired	68	3	Fair	B
75f-15	75f	Impaired	68	3	Fair	B
75e-60	75e	Reference	68	3	Fair	B
75h-69	75h	Impaired	68	3	Fair	B

(Continued)

TABLE F.35 (CONTINUED)
Stream Ratings for Ecoregion 75—Southern Coastal Plain

Station ID	Subecoregion	Condition	Index Score	Numeric Ranking	Narrative Description	Stream Rating
75e-3	75e	Impaired	66	3	Fair	B
75h-1	75h	Impaired	65	3	Fair	B
75j-29	75j	Ref/ Removed	65	3	Fair	B
75e-2	75e	Impaired	65	3	Fair	B
75h-60	75h	Reference	64	4	Poor	C
75h-70	75h	Impaired	64	4	Poor	C
75j-12	75j	Impaired	64	4	Poor	C
75j-11	75j	Impaired	62	4	Poor	C
75j-3	75j	Impaired	61	4	Poor	C
75e-36	75e	Impaired	59	4	Poor	C
75h-66	75h	Reference	59	4	Poor	C
75e-8	75e	Impaired	59	4	Poor	C
75f-77	75f	Impaired	57	4	Poor	C
75e-61	75e	Impaired	56	4	Poor	C
75h-41	75h	Impaired	55	4	Poor	C
75j-3-1	75j	Impaired	50	4	Poor	C
75e-46	75e	Impaired	49	5	Very poor	C
75h-72	75h	Impaired	48	5	Very poor	C
75f-50	75f	Impaired	47	5	Very poor	C
75f-45	75f	Impaired	44	5	Very poor	C

SUBECOREGION 75F—SEA ISLAND FLATWOODS

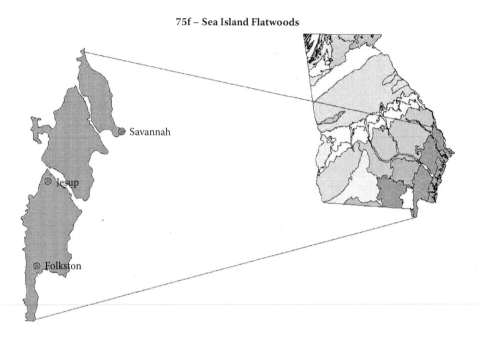

75f – Sea Island Flatwoods

Savannah

Jesup

Folkston

FIGURE F.21   Subecoregion 75f—Sea Island Flatwoods.

The Sea Island Flatwoods are poorly drained flat plains with lower elevations and less dissection than Subecoregion 65l. Pleistocene sea levels rose and fell several times creating different terraces and shoreline deposits. Spodosols and other wet soils are common, although small areas of better drained soils add some ecological diversity. Trail Ridge is in this region, forming the boundary with Subecoregion 75g. Loblolly and slash pine plantations cover much of the region. Water oak, willow oak, sweetgum, blackgum, and cypress occur in wet areas.

TABLE F.36
Characteristic Reference Stream Land Use, Habitat, and Chemistry Data for Subecoregion 75f—Sea Island Flatwoods

	Parameter	Mean	Median	Range
Catchment	% Natural	82.8	81.5	78.5–88.6
Land Use	% Agriculture	0	0	0
	% Silviculture	9.8	12.2	4.8–12.5
	% Urban	7.3	6.5	6.0–9.3
Habitat	Total Habitat Score (200)	153.0	151.0	146–164
	Epifaunal Substrate (20)	16.0	17.5	11–18
	Pool Substrate Characterization (20)	14.0	15.0	10–16
	Pool Variability (20)	11.0	10.0	9–15
	Sediment Deposition (20)	15.3	15.5	10–20
	Channel Flow Status (20)	18.0	19.5	13–20
	Channel Alteration (20)	17.5	17.5	15–20
	Channel Sinuosity (20)	12.3	10.5	9–19
	Bank Stability (L) (10)	8.8	8.5	8–10
	Bank Stability (R) (10)	8.8	8.5	8–10
	Vegetative Protection (L) (10)	8.0	8.5	6–9
	Vegetative Protection (R) (10)	8.3	8.5	6–10
	Riparian Vegetative Width (L) (10)	8.3	9.0	6–9
	Riparian Vegetative Width (R) (10)	7.0	8.0	3–9
In-Stream	% Silt/Clay	51.8	53.5	0–100
Habitat	% Sand	48.3	46.5	0–100.0
(substrate)	% Gravel	0	0	0
	% Cobble	0	0	0
	% Boulder	0	0	0
	% Bedrock	0	0	0
Chemistry	Conductivity (mS/cm)	.117	.120	.051–.179
(in situ)	Dissolved Oxygen (mg/l)	5.7	6.6	3.5–7.1
	pH (SU)	4.6	4.2	3.7–6.0
	Turbidity (NTU)	6.9	3.4	0–17.4
Chemistry	Alkalinity (mg/l as $CaCO_3$)	20.9	20.9	20.9
(laboratory)	Hardness (mg/l as $CaCO_3$)	40.1	40.1	40.1
	Ammonia (mg/l as N)	6.43	6.43	BD–6.43
	Nitrate–Nitrite (mg/l as N)	0.136	0.053	BD–0.315
	Total Phosphorous (mg/l as P)	0.089	0.089	BD–0.113
	Copper (mg/l)	0.003	0.003	BD–0.003
	Iron (mg/l)	0.87	0.96	BD–1.19
	Manganese (mg/l)	0.036	0.036	BD–0.036
	Zinc (mg/l)	0.017	0.017	BD–0.017

Note: BD = Below detection.

FIGURE F.22 Typical reference stream for Subecoregion 75f—Sea Island Flatwoods.

FIGURE F.23 Typical impaired stream for Subecoregion 75f—Sea Island Flatwoods.

TABLE F.37
Discriminating Invertebrate Metrics for Subecoregion
75f —Sea Island Flatwoods

Index 75f

Metric	Metric Category
% Oligochaeta	Composition
% Tanypodinae/Total Chironomidae	
Tolerant Taxa	Tolerance/Intolerance
% Filterer	Functional Feeding Group

TABLE F.38
Descriptive Statistics for Reference Streams in Subecoregion 75f—Sea
Island Flatwoods

Metrics	DE	Minimum	Percentile ($n = 4$)					Maximum
			5th	25th	50th	75th	95th	
% Oligochaeta	0.7	0.0	0.0	0.0	1.9	4.8	7.4	8.1
% Tanypodinae/ Total Chironomidae	0.8	0.0	0.0	0.0	0.0	0.1	0.4	0.5
Tolerant Taxa	0.8	4.0	4.2	4.8	6.5	8.5	9.7	10.0
% Filterer	0.7	0.0	0.0	0.0	0.2	0.6	0.9	0.9

Note: n = Number of reference sites used; DE = discrimination efficiency between reference and impaired conditions.

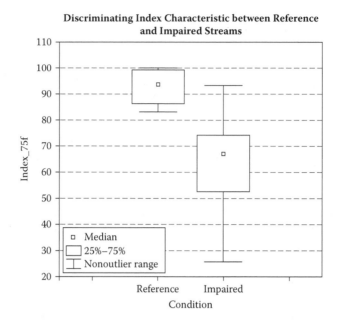

FIGURE F.24 Discriminating index characteristics between reference and impaired streams in Subecoregion 75f—Sea Island Flatwoods.

TABLE F.39

Description of Numeric Rankings for Subecoregion 75f—Sea Island Flatwoods

Index Score	Numeric Ranking	Percentile (n = 15)
98 and above	1	Above 95th
86–97	2	Below 95th, above 75th
60–85	3	Below 75th, above 25th
41–59	4	Below 25th, above 5th
40 and below	5	Below 5th

Note: n = All reference and impaired sites in Subecoregion 75f.

TABLE F.40
Stream Ratings for Subecoregion 75f—Sea Island Flatwoods

Station ID	Condition	Index Score	Numeric Ranking	Narrative Description	Stream Rating
75f-61	Reference	100	1	Very good	A
75f-91	Reference	98	1	Very good	A
75f-28	Impaired	93	2	Good	A
75f-95	Reference	89	2	Good	A
75f-126	Reference	83	3	Fair	B
75f-44	Impaired	79	3	Fair	B
75f-48	Impaired	74	3	Fair	B
75f-70	Impaired	74	3	Fair	B
75f-137	Impaired	69	3	Fair	B
75f-132	Impaired	67	3	Fair	B
75f-15	Impaired	65	3	Fair	B
75f-77	Impaired	55	4	Poor	C
75f-127	Impaired	53	4	Poor	C
75f-50	Impaired	47	4	Poor	C
75f-45	Impaired	26	5	Very poor	C

Appendix G: Summary of Biotic Indices and Precision Measures for Ecoregion 45—Piedmont

TABLE G.1
Biotic Index of Primary Ecoregion 45—Piedmont Developed from the Georgia Ecoregions Project

Metric	Metric Category
Coleoptera Taxa	Richness
% Chironomidae	Composition
% Plecoptera	
% Intolerant Individuals	Tolerance/Intolerance
North Carolina Biotic Index (NCBI)	

TABLE G.2
Average Precision Measure Values of Standardized Metric Scores and Discrimination Efficiencies for Metrics That Comprise the Biotic Index for Primary Ecoregion 45—Piedmont

Metric	RPD	RMSE	CV	DE
Coleoptera Taxa	32.0	25.6	71.8	0.6
% Chironomidae	34.8	28.7	57.9	0.7
% Plecoptera	29.4	12.3	153	0.7
% Intolerant Individuals	67.0	29.1	144	0.6
North Carolina Biotic Index (NCBI)	23.5	29.6	53.4	0.6

Note: RPD = Relative percent difference; RMSE = root mean square of error; CV = coefficient of variability; DE = discrimination efficiency.

TABLE G.3
The Biotic Index of Subecoregion 45a—Southern Inner Piedmont Developed from the Georgia Ecoregions Project

Metric	Metric Category
Plecoptera Taxa	Richness
% Trichoptera	Composition
% Chironomus Cricotopus/TC	
Tolerant Taxa	Tolerance
% Scraper	Functional Feeding Group
Clinger Taxa	Habitat

TABLE G.4
Average Precision Measure Values of Standardized Metric Scores and Discrimination Efficiencies for Metrics That Comprise the Biotic Index for Subecoregion 45a—Southern Inner Piedmont

Metric	RPD	RMSE	CV	DE
Plecoptera Taxa	17.3	20.1	56.7	0.5
% Trichoptera	31.5	37.0	103	0.8
% Chironomus Cricotopus/TC	5.3	9.4	11.2	1.0
Tolerant Taxa	12.2	24.1	35.1	1.0
% Scraper	25.5	28.0	86.7	0.8
Clinger Taxa	10.4	34.2	54.4	0.9

Note: RPD = Relative percent difference; RMSE = root mean square of error; CV = coefficient of variability; DE = discrimination efficiency.

TABLE G.5
Biotic Index of Subecoregion 45b—Southern Outer Piedmont Developed from the Georgia Ecoregions Project

Metric	Metric Category
Coleoptera Taxa	Richness
% Oligochaeta	Composition
% Plecoptera	
Shredder Taxa	Functional Feeding Group
Scraper Taxa	
Swimmer Taxa	Habitat

TABLE G.6
Average Precision Measure Values of Standardized Metric Scores and Discrimination Efficiencies for Metrics That Comprise the Biotic Index for Subecoregion 45b–Southern Outer Piedmont

Metrics	RPD	RMSE	CV	DE
Coleoptera Taxa	44.4	8.6	64.5	0.9
% Oligochaeta	1.2	13.6	15.6	0.8
% Plecoptera	33.3	1.1	245	0.8
Shredder Taxa	49.2	19.4	75.4	0.9
Scraper Taxa	6.7	4.6	18.8	0.9
Swimmer Taxa	83.3	30.8	100	0.9

Note: RPD = Relative percent difference; RMSE = root mean square of error; CV = coefficient of variability; DE = discrimination efficiency.

TABLE G.7
Biotic Index of Subecoregion 45c—Carolina Slate Belt Developed from the Georgia Ecoregions Project

Metric	Metric Category
Tanytarsini Taxa	Richness
% Odonata	Composition
% Tanypodinae/Total Chironomidae	
Dominant Iindividual	Tolerance
% Intolerant Individuals	
% Shredder	Functional Feeding Group
Swimmer Taxa	Habitat

TABLE G.8
Average Precision Measure Values of Standardized Metric Scores and Discrimination Efficiencies for Metrics That Comprise the Biotic Index for Subecoregion 45c—Carolina Slate Belt

Metric	RPD	RMSE	CV	DE
Tanytarsini Taxa	0.0	0.0	0.0	0.6
% Odonata	99.8	35.4	141	0.6
% Tanypodinae/TC	5.4	6.5	7.6	0.6
Dominant Individual	25.6	8.6	36.2	0.6
% Intolerant Individuals	100	13.2	141	0.8
% Shredder	0.0	0.0	0.0	1.0
Swimmer Taxa	100	17.7	141	0.4

Note: RPD = Relative percent difference; RMSE = root mean square of error; CV = coefficient of variability; DE = discrimination efficiency.

TABLE G.9
Biotic Index of Subecoregion 45d—Talladega Upland Developed from the Georgia Ecoregions Project

Metric	Metric Category
Coleoptera Taxa	Richness
% Tanypodinae/Total Chironomidae	Composition
% Odonata	
North Carolina Biotic Index (NCBI)	Tolerance
% Tolerant Individuals	
Shredder Taxa	Functional Feeding Group

TABLE G.10
Average Precision Measure Values of Standardized Metric Scores and Discrimination Efficiencies for Metrics That Comprise the Biotic Index for Subecoregion 45d—Talladega Upland

Metric	RPD	RMSE	CV	DE
Coleoptera Taxa	20.0	8.0	28.3	0.8
% Tanypodinae/TC	0.0	0.0	0.0	1.0
% Odonata	11.8	14.9	16.7	0.8
NCBI	5.1	4.4	22.5	1.0
% Tolerant Individuals	15.9	17.1	7.2	1.0
Shredder Taxa	14.3	16.5	20.2	0.4

Note: RPD = Relative percent difference; RMSE = root mean square of error; CV = coefficient of variability; DE = discrimination efficiency.

TABLE G.11
Biotic Index of Subecoregion 45h—Pine Mountain Ridges Developed from the Georgia Ecoregions Project

Metric	Metric Category
Plecoptera Taxa	Richness
% Ephemeroptera	Composition
% Plecoptera	
% Intolerant Individuals	Tolerance
% Scraper	Functional Feeding Group
% Clinger	Habitat

TABLE G.12
Average Precision Measure Values of Standardized Metric Scores and Discrimination Efficiencies for Metrics That Comprise the Biotic Index for Subecoregion 45h—Pine Mountain Ridges

Metric	RPD	RMSE	CV	DE
Plecoptera Taxa	18.9	33.9	83.0	1.0
% Ephemeroptera	100	32.6	40.9	0.8
% Plecoptera	31.0	40.3	87.1	0.8
% Intolerant Individual	100	27.9	118	0.8
% Scraper	41.0	36.0	64.0	0.6
% Clinger	14.6	19.4	34.0	0.6

Note: RPD = Relative percent difference; RMSE = root mean square of error; CV = coefficient of variability; DE = discrimination efficiency.

SUMMARY OF BIOTIC INDICES AND PRECISION MEASURES FOR ECOREGION 65—SOUTHEASTERN PLAINS

TABLE G.13
Biotic Index of Primary Ecoregion 65—Southeastern Plains Developed from the Georgia Ecoregions Project

Metric	Metric Category
% Coleoptera	Composition
% Oligochaeta	
Intolerant Taxa	Tolerance
% Intolerant Individuals	
% Predator	Functional Feeding Group
% Clinger	Habitat

TABLE G.14
Average Precision Measure Values of Standardized Metric Scores and Discrimination Efficiencies for Metrics That Comprise the Biotic Index for the Primary Ecoregion 65—Southeastern Plains

Metric	RPD	RMSE	CV	DE
% Coleoptera	40.7	30.2	94.1	0.5
% Oligochaeta	5.3	11.6	12.5	0.6
Intolerant Taxa	32.4	33.5	85.7	0.5
% Intolerant Individuals	42.6	30.8	104	0.6
% Predator	23.1	20.4	65.2	0.5
% Clinger	30.3	29.5	81.4	0.5

Note: RPD = Relative percent difference; RMSE = root mean square of error; CV = coefficient of variability; DE = discrimination efficiency.

TABLE G.15
Biotic Index of Subecoregion 65c—Sand Hills Developed from the Georgia Ecoregions Project

Metric	Metric Category
% Trichoptera	Composition
Tolerant Taxa	Tolerance
Intolerant Taxa	
% Scraper	Functional Feeding Group
Clinger Taxa	Habitat

TABLE G.16
Average Precision Measure Values of Standardized Metric Scores and Discrimination Efficiencies for the Metrics That Comprise the Biotic Index for Subecoregion 65c—Sand Hills

Metrics	RPD	RMSE	CV	DE
% Trichoptera	39.5	29.3	76.7	0.7
Tolerant Taxa	6.6	19.7	27.3	0.8
Intolerant Taxa	28.1	39.3	73.7	0.8
% Scraper	58.6	31.5	101	0.9
Clinger Taxa	24.3	28.1	39.2	0.6

Note: RPD = Relative percent difference; RMSE = root mean square of error; CV = coefficient of variability; DE = discrimination efficiency.

TABLE G.17
Biotic Index of Subecoregion 65d–Southern Hilly Gulf Coastal Plain Developed from the Georgia Ecoregions Project

Metric	Metric Category
Plecoptera Taxa	Richness
% Chironomidae	Composition
% Hydropsychidae/EPT	
% Filterer	Functional Feeding Group
Swimmer Taxa	Habitat

TABLE G.18
Average Precision Measure Values of Standardized Metric Scores and Discrimination Efficiencies for Metrics That Comprise the Biotic Index for Subecoregion 65d—Southern Hilly Gulf Coastal Plain

Metric	RPD	RMSE	CV	DE
Plecoptera Taxa	9.5	21.4	32.5	0.7
% Chironomidae	8.6	23.6	29.4	0.7
% Hydropsychidae/ EPT	13.2	12.7	15.8	0.6
% Filterer	43.1	29.7	52.3	0.7
Swimmer Taxa	12.5	35.3	80.7	0.6

Note: RPD = Relative percent difference; RMSE = root mean square of error; CV = coefficient of variability; DE = discrimination efficiency.

TABLE G.19
Biotic Index of Subecoregion 65g—Dougherty Plain Developed from the Georgia Ecoregions Project

Metric	Metric Category
EPT Taxa	Richness
% Oligochaeta	Composition
% Intolerant Individuals	
HBI	Functional Feeding Group
Filterer Taxa	
Clinger Taxa	Habitat

TABLE G.20
Average Precision Measure Values of Standardized Metric Scores and Discrimination Efficiencies for Metrics That Comprise the Biotic Index for Subecoregion 65g—Dougherty Plain

Metric	RPD	RMSE	CV	DE
EPT Taxa	0.0	0.0	0.0	1.0
% Oligochaeta	100	47.1	141	1.0
% Intolerant Individuals	0.0	0.0	0.0	1.0
HBI	15.2	4.1	21.5	1.0
Filterer Taxa	33.3	10.6	47.1	0.8
Clinger Taxa	20.0	7.2	28.3	1.0

Note: RPD = Relative percent difference; RMSE = root mean square of error; CV = coefficient of variability; DE = discrimination efficiency.

TABLE G.21
Biotic Index of Subecoregion 65h—Tifton Upland Developed from the Georgia Ecoregions Project

Metric	Metric Category
Tanytarsini Taxa	Richness
Shannon–Wiener Base e	
% Oligochaeta	Composition
% Tanytarsini	
NCBI	Tolerance
% Predator	Functional Feeding Group
Clinger Taxa	Habitat

TABLE G.22
Average Precision Measure Values of Standardized Metric Scores and Discrimination Efficiencies for Metrics That Comprise the Biotic Index for Subecoregion 65h—Tifton Upland

Metric	RPD	RMSE	CV	DE
Tanytarsini Taxa	64.8	40.9	78.9	0.8
Shannon–Wiener Base e	11.4	13.9	16.3	0.7
% Oligochaeta	7.3	16.5	18.8	0.9
% Tanytarsini	41.4	15.9	59.7	1.0
NCBI	14.5	19.9	27.0	0.8
% Predator	3.4	24.5	31.0	0.6
Clinger Taxa	25.1	28.9	37.7	0.9

Note: RPD = Relative percent difference; RMSE = root mean square of error; CV = coefficient of variability; DE = discrimination efficiency.

TABLE G.23
Biotic Index of Subecoregion 65k—Coastal Plain Red Uplands Developed from the Georgia Ecoregions Project

Metric	Metric Category
% Gastropoda	Composition
% Orthocladiinae/Total Chironomidae	
% Coleoptera	
% Hydropsychidae/Total Trichoptera	
% Filterer	Functional Feeding Group
% Collector	

TABLE G.24
Average Precision Measure Values of Standardized Metric Scores and Discrimination Efficiencies for Metrics That Comprise the Biotic Index for Subecoregion 65k—Coastal Plain Red Uplands

Metric	RPD	RMSE	CV	DE
% Gastropoda	25.0	50.1	83.6	0.8
% Orthocladiinae/Total Chironomidae	63.9	34.0	107	0.6
% Coleoptera	28.2	24.4	82.5	0.6
% Hydropsychidae/ Total Trichoptera	50.0	53.5	107	0.6
% Filterer	27.2	33.4	61.4	0.6
% Collector	13.4	17.6	30.2	0.9

Note: RPD = Relative percent difference; RMSE = root mean square of error; CV = coefficient of variability; DE = discrimination efficiency.

TABLE G.25
Biotic Index of Subecoregion 65l—Atlantic Southern Loam Plains Developed from the Georgia Ecoregions Project

Metric	Metric Category
EPT Taxa	Richness
Diptera Taxa	
% EPT	Composition
% Trichoptera	
HBI	Tolerance
Predator Taxa	Functional Feeding Group
Clinger Taxa	Habitat

TABLE G.26
Average Precision Measure Values of Standardized Metric Scores and Discrimination Efficiencies for Metrics that Comprise the Biotic Index for Subecoregion 65l—Atlantic Southern Loam Plains

Metric	RPD	RMSE	CV	DE
EPT Taxa	75.0	25.2	118	0.8
Diptera Taxa	9.4	21.8	35.2	0.6
% EPT	67.0	49.8	143	0.8
% Trichoptera	66.7	9.0	162	0.9
HBI	19.4	15.5	35.4	0.6
Predator Taxa	8.5	25.6	46.9	0.7
Clinger Taxa	6.7	17.6	73.7	0.8

Note: RPD = Relative percent difference; RMSE = root mean square of error; CV = coefficient of variability; DE = discrimination efficiency.

TABLE G.27
Biotic Index of Subecoregion 65o–Tallahassee Hills/Valdosta Limesink Developed from the Georgia Ecoregions Project

Metric	Metric Category
Chironomidae Taxa	Richness
% Oligochaeta	Composition
Beck's Index	Tolerance
NCBI	
Scraper Taxa	Functional Feeding Group
Burrower Taxa	Habitat
Sprawler Taxa	

TABLE G.28
Average Precision Measure Values of Standardized Metric Scores and Discrimination Efficiencies for Metrics That Comprise the Biotic Index for Subecoregion 65o—Tallahassee Hills/Valdosta Limesink

Metric	RPD	RMSE	CV	DE
Chironomidae Taxa	14.4	22.0	31.7	0.8
% Oligochaeta	3.0	5.6	5.9	0.8
Beck's Index	9.4	26.0	34.4	0.4
NCBI	19.2	23.3	40.2	0.6
Scraper Taxa	0.0	32.5	40.2	0.8
Burrower Taxa	18.3	12.7	31.8	0.6
Sprawler Taxa	11.4	24.8	28.0	0.8

Note: RPD = Relative percent difference; RMSE = root mean square of error; CV = coefficient of variability; DE = discrimination efficiency.

SUMMARY OF BIOTIC INDICES AND PRECISION
MEASURES FOR ECOREGION 66—BLUE RIDGE

TABLE G.29
Biotic Index of Primary Ecoregion 66—Blue Ridge
Developed from the Georgia Ecoregions Project

Metric	Metric Category
Plecoptera Taxa	Richness
Simpson's Index	
% Trichoptera	Composition
% Intolerant Individuals	
NCBI	Tolerance
Predator Taxa	Functional Feeding Group
Burrower Taxa	Habitat

TABLE G.30
Average Precision Measure Values of Standardized Metric Scores and
Discrimination Efficiencies for Metrics That Comprise the Biotic Index For
Primary Ecoregion 66—Blue Ridge

Metric	RPD	RMSE	CV	DE
Plecoptera Taxa	31.2	25.5	46.2	0.7
Simpson's Index	20.4	26.1	37.5	0.6
% Trichoptera	14.9	24.5	35.1	0.5
% Intolerant Individuals	24.2	27.9	48.5	0.6
NCBI	22.7	31.2	49.3	0.7
Predator Taxa	15.7	21.4	28.8	0.8
Burrower Taxa	21.4	22.7	30.1	0.5

Note: RPD = Relative percent difference; RMSE = root mean square of error; CV = coefficient of variability; DE = discrimination efficiency.

TABLE G.31
Biotic Index of Subecoregion 66d—Southern Crystalline Ridges and Mountains Developed from the Georgia Ecoregions Project

Metric	Metric Category
Diptera Taxa	Richness
% Plecoptera	Composition
% Odonata	
% Dominant Individuals	Tolerance
% Shredder	Functional Feeding Group
Clinger Taxa	Habitat

TABLE G.32
Average Precision Measure Values of Standardized Metric Scores and Discrimination Efficiencies for Metrics That Comprise the Biotic Index for Subecoregion 66d—Southern Crystalline Ridges and Mountains

Metric	RPD	RMSE	CV	DE
Diptera Taxa	4.1	18.5	26.8	0.8
% Plecoptera	20.2	10.8	24.3	0.6
% Odonata	50.7	42.3	72.6	1.0
% Dominant Individuals	3.8	18.3	28.6	0.6
% Shredder	8.6	15.5	42.8	0.8
Clinger Taxa	13.8	16.8	20.4	0.6

Note: RPD = Relative percent difference; RMSE = root mean square of error; CV = coefficient of variability; DE = discrimination efficiency.

TABLE G.33
Biotic Index of Subecoregion 66g—Southern Metasedimentary Mountains Developed from the Georgia Ecoregions Project

Metric	Metric Category
EPT Taxa	Richness
% Chironomidae	Composition
% Tanypodinae/Total Chironomidae	
NCBI	Tolerance
% Dominant Individuals	
Scraper Taxa	Functional Feeding Group

TABLE G.34
Average Precision Measure Values of Standardized Metric Scores and Discrimination Efficiencies for Metrics That Comprise the Biotic Index for Subecoregion 66g—Southern Metasedimentary Mountains

Metric	RPD	RMSE	CV	DE
EPT taxa	5.4	19.2	27.6	0.9
% Chironomidae	35.9	22.4	54.6	0.9
% Tanypodinae/Total Chironomidae	35.4	20.4	76.4	0.9
NCBI	41.8	37.8	59.6	0.7
% Dominant Individuals	39.4	35.8	52.8	0.7
Scraper Taxa	13.0	24.5	40.8	0.9
% Clinger	4.9	14.7	24.9	0.7

Note: RPD = Relative percent difference; RMSE = root mean square of error; CV = coefficient of variability; DE = discrimination efficiency.

TABLE G.35
Biotic Index of Subecoregion 66j—Broad Basins Developed from the Georgia Ecoregions Project

Metric	Metric Category
Simpson's Diversity Index	Richness
Margalef's Index	
% Tanytarsini	Composition
% Intolerant Individuals	Tolerance
Predator Taxa	Functional Feeding Group
Sprawler Taxa	Habitat

TABLE G.36
Average Precision Measure Values of Standardized Metric Scores and Discrimination Efficiencies for Metrics That Comprise the Biotic Index for Subecoregion 66j—Broad Basins

Metric	RPD	RMSE	CV	DE
Simpson's Diversity Index	38.9	12.6	59.3	0.8
Margalef's Index	10.4	34.0	14.9	0.8
% Tanytarsini	63.5	34.2	80.9	0.8
% Intolerant Individuals	29.9	20.0	36.8	0.6
Predator Taxa	16.1	21.6	32.9	0.8
Sprawler Taxa	19.2	20.0	32.0	0.6

Note: RPD = Relative percent difference; RMSE = root mean square of error; CV = coefficient of variability; DE = discrimination efficiency.

SUMMARY OF BIOTIC INDICES AND PRECISION MEASURES FOR ECOREGION 67—RIDGE AND VALLEY AND ECOREGION 68—SOUTHWESTERN APPALACHIANS

TABLE G.37
Biotic Index of Primary Ecoregion 67—Ridge and Valley Developed from the Georgia Ecoregions Project

Metric	Metric Category
EPT Taxa	Richness
Plecoptera Taxa	
% Plecoptera	Composition
% Isopoda	
HBI	Tolerance
Clinger Taxa	Habitat

TABLE G.38
Average Precision Measure Values of Standardized Metric Scores and Discrimination Efficiencies for Metrics That Comprise the Biotic Index for Primary Ecoregion 67—Ridge and Valley

Metric	RPD	RMSE	CV	DE
EPT Taxa	17.4	23.1	42.1	0.8
Plecoptera Taxa	26.3	35.6	85.1	0.8
% Plecoptera	38.5	25.4	166	0.8
% Isopoda	4.8	11.6	12.6	0.7
HBI	8.6	11.3	16.2	0.8
Clinger Taxa	7.8	21.4	30.4	0.7

Note: RPD = Relative percent difference; RMSE = root mean square of error; CV = coefficient of variability; DE = discrimination efficiency.

TABLE G.39
Biotic Index of Subecoregion 67f&i—Southern Limestone/ Dolomite Valleys and Low Rolling Hills Developed from the Georgia Ecoregions Project

Metric	Metric Category
EPT Taxa	Richness
Plecoptera Taxa	
% EPT	Composition
NCBI	Tolerance
Scraper Taxa	Functional Feeding Group
% Clinger	Habitat

TABLE G.40
Average Precision Measure Values of Standardized Metric Scores and Discrimination Efficiencies for the Metrics That Comprise the Biotic Index for Subecoregion 67f&i—Southern Limestone/Dolomite Valleys and Low Rolling Hills

Metric	RPD	RMSE	CV	DE
EPT Taxa	17.1	20.6	31.7	1.0
Plecoptera Taxa	23.1	40.3	91.8	1.0
% EPT	19.7	28.6	48.5	1.0
NCBI	7.6	12.6	13.9	0.8
Scraper Taxa	18.3	28.6	35.0	0.8
% Clinger	8.7	19.1	25.0	1.0
Clinger Taxa	10.9	19.2	28.1	0.7

Note: RPD = Relative percent difference; RMSE = root mean square of error; CV = coefficient of variability; DE = discrimination efficiency.

TABLE G.41
Biotic Index of Subecoregion 67h—Southern Sandstone Ridges Developed from the Georgia Ecoregions Project

Metric	Metric Category
Plecoptera Taxa	Richness
% Gastropoda	Composition
% Tolerant Individuals	Tolerance
HBI	
Scraper Taxa	Functional Feeding Group
Swimmer Taxa	Habitat

TABLE G.42
Average Precision Measure Values of Standardized Metric Scores and Discrimination Efficiencies for Metrics That Comprise the Biotic Index for Subecoregion 67h—Southern Sandstone Ridges

Metric	RPD	RMSE	CV	DE
Plecoptera Taxa	55.6	47.8	78.6	0.5
% Gastropoda	0.0	0.0	0.0	1.0
% Tolerant Individuals	19.5	19.7	27.5	1.0
HBI	0.6	0.3	0.9	1.0
Scraper Taxa	20.0	12.6	28.3	1.0
Swimmer Taxa	33.3	35.4	47.1	1.0

Note: RPD = Relative percent difference; RMSE = root mean square of error; CV = coefficient of variability; DE = discrimination efficiency.

TABLE G.43
Biotic Index of Primary Ecoregion 68—Southwestern Appalachians Developed from the Georgia Ecoregions Project

Metric	Metric Category
Plecoptera Taxa	Richness
% Hydropsychidae/Total Trichoptera % Tanypodinae/Total Chironomidae	Composition
NCBI	Tolerance
Scraper Taxa	Functional Feeding Group
% Clinger	Habitat

TABLE G.44
Average Precision Measure Values of Standardized Metric Scores and Discrimination Efficiencies for the Metrics That Comprise the Biotic Index for Primary Ecoregion 68—Southwestern Appalachians

Metric	RPD	RMSE	CV	DE
Plecoptera Taxa	0.0	0.0	0.0	0.8
% Hydropsychidae/Total Trichoptera	7.8	6.0	11.1	0.6
% Tanypodinae/Total Chironomidae	47.2	9.0	66.7	0.8
NCBI	1.8	2.8	4.5	1.0
Scraper Taxa	33.3	31.7	47.1	0.8
% Clinger	1.0	1.3	1.4	0.6

Note: RPD = Relative percent difference; RMSE = root mean square of error; CV = coefficient of variability; DE = discrimination efficiency.

SUMMARY OF BIOTIC INDICES AND PRECISION MEASURES FOR ECOREGIONS 75—SOUTHERN COASTAL PLAIN

TABLE G.45
Biotic Index of Primary Ecoregion 75—Southern Coastal Plain Developed from the Georgia Ecoregions Project

Metric	Metric Category
% Noninsect % Oligochaeta % Odonata % Tanypodinae/Total Chironomidae	Composition
Tolerant Taxa % Tolerant Individuals	Tolerance

TABLE G.46
Average Precision Measure Values of Standardized Metric Scores and Discrimination Efficiencies for the Metrics That Comprise the Biotic Index for Primary Ecoregion 75—Southern Coastal Plain

Metric	RPD	RMSE	CV	DE
% Noninsect	22.2	32.2	62.9	0.6
% Oligochaeta	6.0	11.1	12.2	0.7
% Odonata	2.5	9.1	9.8	0.5
% Tanypodinae/Total Chironomidae	3.2	8.1	8.7	0.5
Tolerant Taxa	24.1	32.4	62.5	0.6
% Tolerant Individuals	24.6	30.7	59.6	0.5

Note: RPD = Relative percent difference; RMSE = root mean square of error; CV = coefficient of variability; DE = discrimination efficiency.

TABLE G.47
Biotic Index of Subecoregion 75e—Okefenokee Plains developed from the Georgia Ecoregions Project

Metric	Metric Category
% Noninsect	Composition
% Oligochaeta	
% Isopoda	
% Odonata	
% Tolerant Individuals	Tolerance
% Filterer	Functional Feeding Group

TABLE G.48
Average Precision Measure Values of Standardized Metric Scores and Discrimination Efficiencies for Metrics That Comprise the Biotic Index for Subecoregion 75e—Okefenokee Plains

Metric	RPD	RMSE	CV	DE
% Noninsect	18.1	33.6	63.1	0.9
% Oligochaeta	8.2	23.0	28.8	0.9
% Isopoda	28.4	30.2	50.4	0.6
% Odonata	20.2	21.5	27.1	0.6
% Tolerant Individuals	28.0	15.9	62.0	0.6
% Filterer	41.6	31.8	49.8	0.6

Note: RPD = Relative percent difference; RMSE = root mean square of error; CV = coefficient of variability; DE = discrimination efficiency.

TABLE G.49
Biotic Index of Subecoregion 75f—Sea Island Flatwoods
Developed from the Georgia Ecoregions Project

Metric	Metric Category
% Oligochaeta	Composition
% Tanypodinae/Total Chironomidae	
Tolerant Taxa	Tolerance
% Filterer	Functional Feeding Group

TABLE G.50
Average Precision Measure Values of Standardized Metric Scores and Discrimination Efficiencies for the Metrics That Comprise the Biotic Index for Subecoregion 75f—Sea Island Flatwoods

Metric	RPD	RMSE	CV	DE
% Oligochaeta	10.3	11.6	14.5	0.7
% Tanypodinae/Total Chironomidae	1.9	2.7	2.7	0.8
Tolerant Taxa	50.0	44.0	70.7	0.8
% Filterer	0.0	0.0	0.0	0.7

Note: RPD = Relative percent difference; RMSE = root mean square of error; CV = coefficient of variability; DE = discrimination efficiency.

TABLE G.51
Biotic Index of Subecoregion 75h—Bacon Terraces
Developed from the Georgia Ecoregions Project

Metric	Metric Category
% Oligochaeta	Composition
% Tolerant Individuals	Tolerance
HBI	
% Shredder	Functional Feeding Group
Collector Taxa	
% Filterer	

TABLE G.52

Average Precision Measure Values of Standardized Metric Scores and Discrimination Efficiencies for the Metrics That Comprise the Biotic Index for Subecoregion 75h—Bacon Terraces

Metric	RPD	RMSE	CV	DE
% Oligochaeta	2.4	3.4	3.5	0.8
% Tolerant Individuals	9.8	12.7	13.9	0.8
HBI	27.7	29.3	39.2	0.5
% Shredder	100	15.8	141	0.8
Collector Taxa	4.3	4.5	6.1	0.5
% Filterer	5.2	6.5	7.3	0.5

Note: RPD = Relative percent difference; RMSE = root mean square of error; CV = coefficient of variability; DE = discrimination efficiency.

TABLE G.53

Biotic Index of Subecoregion 75j–Sea Islands/Coastal Marshes Developed from the Georgia Ecoregions Project

Metric	Metric Category
% Noninsect	Composition
% Oligochaeta	
% Tolerant Individuals	Tolerance
Shredder Taxa	Functional Feeding Group

TABLE G.54

Average Precision Measure Values of Standardized Metric Scores and Discrimination Efficiencies for the Metrics That Comprise the Biotic Index for Subecoregion 75j—Sea Islands/Coastal Marshes

Metric	RPD	RMSE	CV	DE
% Noninsect	38.0	37.6	48.9	0.6
% Oligochaeta	7.5	13.0	75.6	0.5
% Tolerant Individuals	35.4	34.1	53.9	0.6
Shredder Taxa	33.3	20.4	30.6	0.5

Note: RPD = Relative percent difference; RMSE = root mean square of error; CV = coefficient of variability; DE = discrimination efficiency.

Index